How to Cheat with Statistics — and Get Away with It

From Data Snooping over Kitchen Sink
Regression to "Creative Reporting"

How to Cheat with Statistics — and Get Away with It

From Data Snooping over Kitchen Sink Regression to "Creative Reporting"

Gunter Meissner

Columbia University, USA
Derivatives Software

World Scientific

NEW JERSEY · LONDON · SINGAPORE · BEIJING · SHANGHAI · HONG KONG · TAIPEI · CHENNAI · TOKYO

Published by

World Scientific Publishing Co. Pte. Ltd.

5 Toh Tuck Link, Singapore 596224

USA office: 27 Warren Street, Suite 401-402, Hackensack, NJ 07601

UK office: 57 Shelton Street, Covent Garden, London WC2H 9HE

Library of Congress Cataloging-in-Publication Data
Names: Meissner, Gunter, 1957–   author.
Title: How to cheat with statistics — and get away with it : from data snooping over
    kitchen sink regression to "creative reporting" / Gunter Meissner.
Description: New Jersey : World Scientific, [2022] | Includes bibliographic references.
Identifiers: LCCN 2022004405 | ISBN 9789811251719 (hardcover) |
    ISBN 9789811252488 (paperback) | ISBN 9789811251726 (ebook for institutions) |
    ISBN 9789811251733 (ebook for individuals)
Subjects: LCSH: Mathematical statistics. | Statistics--Methodology.
Classification: LCC QA276 .M3985 2022 | DDC 001.4/22--dc23/eng20220328
LC record available at https://lccn.loc.gov/2022004405

British Library Cataloguing-in-Publication Data
A catalogue record for this book is available from the British Library.

For any available supplementary material, please visit
https://www.worldscientific.com/worldscibooks/10.1142/12708#t=suppl

Typeset by Stallion Press
Email: enquiries@stallionpress.com

Preface

During my 30 years of teaching statistics, I came across many weaknesses and flaws in statistics. These weaknesses allow a malevolent researcher to manipulate the inputs, the calculations, and the reporting of results to derive a desired outcome. Somewhat contradictory to the catchy title, "How to Cheat in Statistic and Get Away with It" (marketing is everything ☺), I actually explain how to identify and catch statistical cheaters. So the book should be valuable to everyone who wants to gain a deeper understanding of the weaknesses in statistics and learn how to evaluate statistical research to catch a statistical cheater!

Target Audience

This book should be of interest to everyone interested in statistics. Chapter 1 on Data Input Manipulation and Chapter 3 on Creative Reporting can be understood without any prior knowledge of statistics. Chapter 2 on Manipulating the Statistical Calculations does use some math. But the math is explained in simple terms and should be easy to follow. In addition, the book comes with 18 Excel spreadsheets which should help understanding the math. There are also 7 Python codes, which can be run online at repl.it. For those who want to get deeper into stats, there are several proofs, which can be accessed online (links are in the book).

The book comes with question and problems at the end of each chapter, which should facilitate the usage in a classroom. Answers to the questions and problems are available to instructors, please email Gunter@dersoft.com.

I welcome feedback! If you have any suggestions or comments, please email me at Gunter@dersoft.com.

About the Author

After a lectureship in mathematics and statistics at the Economic Academy Kiel, Gunter Meissner PhD, joined Deutsche Bank in 1990, trading interest rate futures, swaps, and options in Frankfurt and New York. He became Head of Product Development in 1994, responsible for originating algorithms for new derivatives products, which at the time were Lookback Options, Multi-asset Options, Quanto Options, Average Options, Index Amortizing Swaps, and Bermuda Swaptions. In 1995/1996, Gunter Meissner was Head of Options at Deutsche Bank Tokyo. From 1997 to 2007, Gunter was Professor of Finance at Hawaii Pacific University and from 2008 to 2013 Director of the Master in Financial Engineering Program at the University of Hawaii. Currently, he is President of Derivatives Software (www.dersoft.com) and Adjunct Professor of Mathematical Finance at Columbia University and NYU.

Gunter Meissner has published numerous papers and six other books on derivatives and risk management, and is a frequent speaker on conferences and seminars. He can be reached at gunter@dersoft.com. His CV is at www.dersoft.com/cv.pdf.

Contents

Chapter 1

Input Manipulation

In this chapter, we discuss how inputs can be manipulated to derive desired research results. In particular, we analyze how to manipulate input data and the research time frame.

1.1. Manipulating Input Data

Let's first discuss manipulating input data. We will discuss five possible ways: Data Snooping, Biased Sampling, Eliminating Outliers, Using False Inputs, and Data Transformation. Let's start with one of the most applied input data manipulations.

1.1.1. *Data Snooping*

There are researchers with a Machiavellian attitude: The objective, a publication, justifies the means, statistical cheating. One form of statistical cheating is data snooping.

What is Data Snooping, also called Data Fishing, Data Dredging, or Data Butchery?

> *Data Snooping is selecting Data for which the Research Hypothesis works*

In ethical statistical research, first a research hypothesis is formulated. Then the hypothesis is tested with real-world data. Performing a data analysis first and selecting specific data for which the hypothesis works is Data Snooping and constitutes statistical cheating. Here are some examples of Data Snooping.

Example 1.1: *Dogmatic Anti-vaxxer Joel (who is also not vacci-nated against Covid 19 ☹), wants to prove that vaccination causes Autism. He conducts a study with 10,000 vaccinated children and 10,000 non-vaccinated children from 100 countries. Joel does not find an overall positive association between the vaccinated children and Autism. However, he finds that a positive association between vaccination and Autism exists (by chance) in one country. He uses the data from this one country as input data and concludes that vaccination causes Autism.*

Example 1.2: *PhD student Hans wants to prove that smoking does not cause cancer (since his research is funded by the tobacco industry ☺). His data pool of 1,000 smokers and 1,000 non-smokers shows that a significant number of smokers develop cancer in later years. Therefore he conducts his research for individuals under 30. Since few individuals under 30 develop cancer, the research concludes that smoking does not cause cancer.*

Example 1.3: *The National Association of Independent Schools (NAIS) wants to prove that in the US private school children have higher SAT scores than public school children.[1] An NAIS researcher investigates 1,000 private schools and 1,000 public schools in the US. He does not find any association between private schools and higher SAT scores. However, he finds that a certain association between private schools and higher SAT scores exists (by chance) in one state. He uses data from this one state as input data and concludes that private schools have higher SAT scores.*

How to Catch Data Snoopers

That's an easy one! We have to perform out-of-sample testing, also called cross-validation. In rigorous statistical testing, the data set is split into a training set and an out-of-sample set, also called validation set. In Example 1.1, the training set was just one country.

---

[1] Most studies show little evidence that SAT scores of private schools are higher. This study also suffers from a Reporting Bias: Private schools with bad SAT scores simply do not report their scores, inflating private school SAT scores. We discuss the Reporting bias in detail in Chapter 3, Sections 3.1.2 and 3.1.3.

If we test the hypothesis that vaccines cause Autism in the validation set of 99 other countries, we will realize that the hypothesis cannot be generalized and is therefore false.

The same logic applies to Example 1.2: Testing the hypothesis that smoking causes cancer with a validation set of individuals over 30, we will most likely find a significant association between smoking and cancer.

The same rationale applies to Example 1.3: The training set was just one state. Testing the hypothesis that private schools have better SAT scores than public schools out-of-sample in the 49 other states will show that the hypothesis is false.

1.1.2. *Biased Sampling*

Biased sampling, also called convenience sampling, selection bias, or ascertainment bias (particularly in Biology), is another problem with statistical analyses. We can define it as follows:

> *Biased Sampling is collecting data from a non-representative group of the population*

Types of Biased Sampling

Numerous types of Biased Sampling exist. Let's discuss the most critical ones.

Voluntary versus Non-voluntary Biased Sampling

The term "Bias" typically has a negative connotation. Following this interpretation, a researcher may deliberately sample from a non-representative segment or the population to derive biased results. An example would be a study with the intent to show Trump support in the US. A big group of Trump supporters are white males without a college education. We find a big segment of this population at the annual Biker rally in South Dakota, which is pretty cool and attracts about 500,000 bikers annually. Researching Trump support at the annual biker rally would show a higher degree of support for Trump compared to the entire US population and would therefore be biased.

Vice versa, researching Trump support at a liberal University would likely show very little support for Trump, which is also a misleading result if the objective is deriving the degree of Trump support in the entire US.

Biased sampling can also be non-voluntary. In the early days of polling, in 1936, the US Literary Digest magazine predicted that the Republican presidential candidate Alf Landon would beat the incumbent FDR by a wide margin. However, the sample voters consisted mainly of readers of the magazine, who were mostly wealthy Americans, leaning towards a Republican president. FDR won in a landslide.

> *"With careful and prolonged planning, we may reduce or eliminate many potential sources of bias, but seldom will we be able to eliminate all of them."*
>
> (Good and Harding 2006)

Total versus Partial Biased Sampling

Total Biased Sampling occurs when there is no overlap between the sample and the population. This is very rare though. When you conduct a poll for the US presidency, you don't sample in Japan. However, a partial sampling error is very common since a sample is typically not a perfect representation of the population. Statistical adjustments can be made to correct for the sampling error.

For example, a researcher wants to find the average height of the US adult population. For simplicity, let's assume there are 50% adult males and 50% adult females in the US[2] and the average height of males is 5.9 feet and the average height of females is 5.4 feet, following data from the Center for Disease Control. The researcher sends out questionnaires to 1,000 men and 1,000 women. 600 women respond and report an average height of 5.4 feet and 400 men respond reporting an average height of 5.9 feet. To correct for the female/male ratio sample error, we simply apply weighting

---

[2]The male to female birth ratio in the US is about 105/100. However, the life expectancy of females (at birth) is 81.1 years, and the life expectance of males (at birth) is 78.5. So there are actually more female adults than male adults in the US, the ratio being 100/97.95.

factors, i.e., the average height in the US adult population derived from the sample is $(600 \times 5.4 \times 1.6667 + 400 \times 5.9 \times 2.5)/1{,}000/2 = 5.65$ feet.

1.1.3. *Eliminating Outliers*

In the following, we will discuss how to deal with outliers, in particular, if it is legitimate to eliminate them.

Truncating the Data Pool

When I ask my students what to do with outliers, I typically get the answer: "Get rid of them!" This is not the correct answer. Outliers, unless they are erroneous, are part of the data pool. Just eliminating them constitutes data manipulation, which is statistical cheating. Let's show this in two examples.

Example 1.4: *The marketing team of Mephisto LLC wants to prove that more advertising leads to more sales. They collect annual data from the last 12 years and get the following graph:*

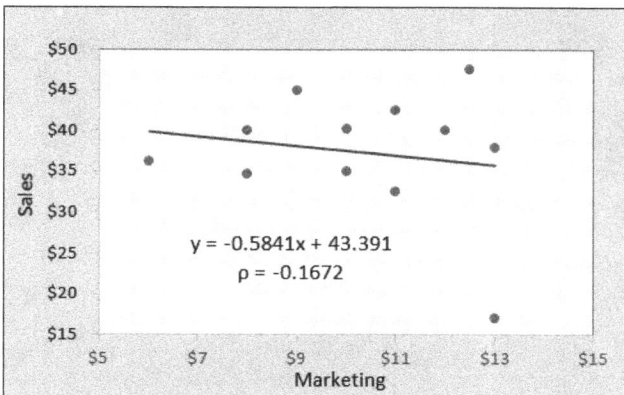

Figure 1.1: Relationship between marketing and sales of Mephisto LLC; data in $ million.

From Figure 1.1, we observe a negative relationship between marketing and sales, so on average the higher the marketing, the lower the sales. We can see this simply by the negative slope of the linear regression function. Statistically the regression coefficient, which measures the slope of the regression function, is −0.5841, as seen in Figure 1.1. The correlation coefficient ρ, which measures the strength and direction of a relationship, is −0.1672. Both β and ρ verify the negative relationship between marketing and sales. We will discuss β and ρ in detail in Chapter 2.

Understandably the Mephisto marketing team is not happy with the outcome of Figure 1.1. They decide to truncate the data by eliminating annual sales under $20 million. This removes the outlier (13, 17). Without the outlier, Figure 1.2 is derived.

From Figure 1.2, we see that the objective is achieved: The relationship between marketing and sales is now positive with a regression coefficient $\beta = 0.6173$ and a correlation coefficient $\rho = 0.2905$.

We have a critical question: Is this truncating of input data, which can eliminate outliers, legitimate or is it statistical cheating? The answer is easy. It is cheating! As stated above, if outliers are not erroneous, they are part of the input data set and have to be included in the statistical analysis!

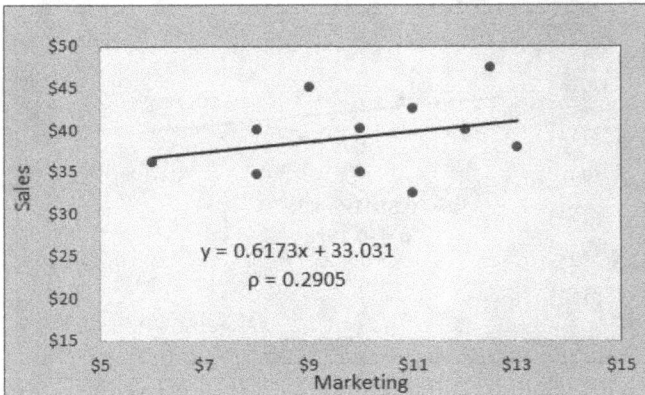

Figure 1.2: Relationship between marketing and sales of Mephisto LLC excluding annual sales data under $20 million; data in $ million.

However, a significant outlier can strongly influence the regression output as in our example. A remedy is running the regression with and without the outlier, which we did with Figures 1.1 and 1.2. This gives the observer a good perception of the relationship between two variables with and without outliers.

Applying Median Not Mean

In Example 1.4, we had two variables, sales, and marketing. Let's look at an example with only one variable, the return of a hedge fund.

Example 1.5: *An investor is considering investing in the hedge fund "Doomed". The investor wants to know the average return (i.e., profit) of the last 7 years. Ordered from lowest to highest, the annual hedge fund return HFR was*

$$HFR = \{-40\%, -20\%, 6\%, 9\%, 10\%, 12\%, 14\%\}$$

The arithmetic mean of the return is

$$(-40\% - 20\% + 6\% + 9\% + 10\% + 12\% + 14\%)/7 = -1.29\%.$$

The hedge fund however reports the median to the investor. A median is the middle value of an ordered number series. So, in the above example, the median is 9%, which the hedge fund reports to the investor as the average performance of the last 7 years.

Is this statistical cheating? Not an easy question. Statistically, the arithmetic mean and the median (and other parameters such as mode or geometric average) are all considered forms of an average. So if the investor did not specifically ask for the arithmetic mean, the hedge is not statistically cheating when reporting the median.

However, as we can see from the derivation of the median, the median is mathematically NOT an average. It is just the middle value of a number series. So reporting the median as an average can be highly misleading.

Our example shows an important fact: Medians ignore one or more outliers! Some statisticians consider this a desired property since ignoring outliers may be a better way to express the central tendency of the number series. In our example, the hedge fund produced profits in 5 out of the 7 years, so the hedge fund may argue that the median of 9% is representative of the average performance.

However, as the author of this book, I advise caution with respect to the median. Unless there is a high probability that the outlier(s) are erroneous, outliers are part of the input data set and should not be ignored! So whenever you see a study reporting the median, which is not uncommon, be aware! And calculate additionally the arithmetic mean. The difference may tell a story.

1.1.4. *Just Use False Inputs: The Hwang Woo-suk Cloning Scandal*

In 2004 and 2005, the Korean Professor of Biotechnology Hwang Woo-suk published two papers in the journal *Science*, reporting breakthroughs in the cloning of human embryos. He claimed to have succeeded in creating human embryonic stems cells with a technology called cell nuclear transfer, a method also used in Dolly the Sheep: The nucleus of a donor egg is removed to create a nucleus-free egg. This egg is then fused with the DNA of the human being to be cloned using electricity.

Sounds all very exciting and at the time it was. Hwang Woo-suk was called the Pride of Korea and given free first-class travel by Korean Airlines. There was only one problem: The results were fabricated. The DNA profiles of Hwang Woo-suk's clones were identical and therefore had to come from the same source. Ja Min Koo, a coauthor, admitted that she had donated oocytes, a germ cell involved in reproduction.

In 2009, Hwang was found guilty of bioethical violations and embezzlement and sentenced to 2 years suspended prison sentence.

What to learn from this? Well, a candid way of cheating is just using false input data. While this is not a cheating method using statistical methodologies, it is a simple way to fool referees. Hence also inputs should be carefully scrutinized when reviewing research. And in the end, we gratefully acknowledge: Cheaters typically get caught!

1.1.5. *Data Transformation: Is Standardization or Transformation to Percentages Statistical Cheating?*

Transforming data is a good thing and a bad thing. The bad thing is that often information of the original data distribution is lost. However, the benefit is that transformed data is often easier to work with.

Sometimes data transformation to different scales is actually necessary. If we want to compare temperatures in Fahrenheit and Celsius, we have to convert degrees from the ridiculous Fahrenheit scale to Celsius or vice versa. The equation is Celsius = (Fahrenheit − 32) × (5/9), or Fahrenheit = 9/5 Celsius + 32.

And the failure of data transformation can be embarrassing: In 1999, the Mars orbiter was destroyed by Mars' atmosphere because scientists failed to transfer inches to metric units! That's what you get when don't want to convert to the metric scale ☺.

When transforming data is performed voluntarily, we have to make sure that the benefits outweigh the limitations. Let's discuss several forms of voluntary data transformation and discuss if this is statistical cheating or legitimate.

Standardization of Data

In statistics, we can define standardization, also called *z*-score normalization, as

> *Standardization is the transformation of an original distribution to a distribution with a mean of 0 and a standard deviation of 1.*

It's about time that we use some equations: For a normal distribution N, we can formalize standardization as

$$N(\mu, \sigma) \rightarrow N(0, 1). \tag{1.1}$$

Equation (1.1) reads: The normal distribution N with a mean of μ and a standard deviation of σ is transformed into a standard normal distribution with a mean of 0 and a standard deviation of 1.

How do we transform a distribution to having a 0 mean and a standard deviation of 1? We use the famous z-score. It is

$$z = \frac{x - \mu}{\sigma}, \tag{1.2}$$

where x can be a scalar (a single number) or a vector (a number series). Let's first look at transforming a scalar.

Transforming a Scalar With the z-Score

Transforming a scalar with the z-score is best explained with an example.

Example 1.6: *Overachieving high school student Jake takes a math prep exam on a scale of 0 to 100 to prepare for the math SATs which has a scale of 200 to 800.*[3] *Jake scores 70 on the prep exam and 700 on the SAT. On which exam did Jake do better relative to the other test-takers?*

We can use the z-score to compare the results. We do need the mean and standard deviation of the scores of all students of each test. This is given in Table 1.1.

In conclusion, Jake did better on the SATs than the Prep exam relative to other test-takers. So it was worth studying for and taking the prep exam!

---

[3]The SAT math (and verbal) scale of 200 to 800 seems odd, to use British English. But there is some logic. If a test-takers has no knowledge at all, they will on average get 25% of answers correct by guessing, since there are 4 answers to select from (and no penalty for wrong answers). So they would score 200 out of 800. However, a test-taker can theoretically get less than 25% correct. In this case, their SAT score will still be 200, since this is the minimum score set by the SAT. The SAT designers could also have used a scale of 0 to 100. However, test-takers with little knowledge would score a numerically low absolute number, which could be disillusioning.

Table 1.1: Jake's *z*-score for the Prep exam and SAT, derived with Equation (1.2).

	Prep exam	SAT
Jake's test score x	70	700
Test score mean μ	55	530
Test score standard deviation σ	20	200
Jake's *z*-score	**0.75**	**0.85**

Figure 1.3: Normal distribution equation and Carl Friedrich Gauss on the former 10 Deutsch Mark bill.

Transforming a Vector With the z-Score

Let's assume we have a normally distributed variable x, with a mean of μ and a standard deviation of σ. Carl-Friedrich Gauss (1777–1855) came up with the equation, called the Gaussian curve or bell curve:

$$f(x; \mu, \sigma) = \frac{1}{\sigma\sqrt{2\pi}} e^{-\frac{1}{2}\left(\frac{x-\mu}{\sigma}\right)^2}. \tag{1.3}$$

This equation and a picture of Gauss were actually on the 10 Deutsch Mark bill. We Germans are very proud of our math heroes ☺ (see Figure 1.3).

Equation (1.3) with $\mu = -1$ and $\sigma = 2$, is displayed in Figure 1.4.

Let's now transform the normal distribution of Equation (1.3) with $\mu = -1$ and $\sigma = 2$ to a standard normal distribution with $\mu = 0$ and $\sigma = 1$. This can be done easily with the z-score of Equation (1.2). We simply deduct $\mu = -1$ from every value x and divide it by $\sigma = 2$.

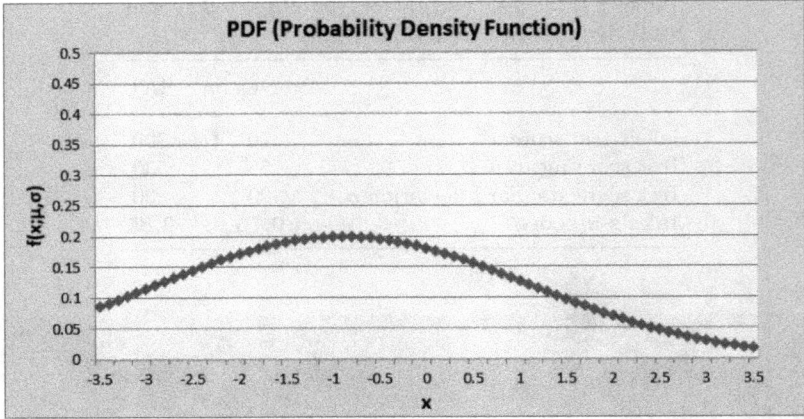

Figure 1.4: The PDF of a normal distribution with $\mu = -1$ and $\sigma = 2$.

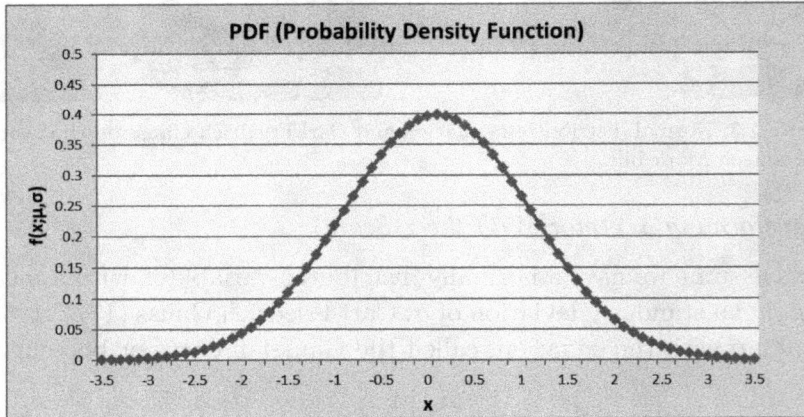

Figure 1.5: The PDF of a standard normal distribution, i.e., $\mu = 0$ and $\sigma = 1$.

Doing so, we get nice simple equation, which is the standard normal distribution:

$$f(x; 0, 1) = \frac{1}{\sqrt{2\pi}} e^{-\frac{1}{2}x^2}. \tag{1.4}$$

We now have derived the well-known standard normal distribution, displayed in Figure 1.5.

For a spreadsheet deriving the normal distribution with different values for μ and σ, see www.dersoft.com/normaldist.xlsx.

The z-score transformation of a vector x works well if x is normally or at least approximately normally distributed. This is because the z-score only uses the first 2 moments of a distribution, the mean μ and the standard deviation σ in the transformation and the standard normal distribution is fully characterized by the first two moments. The third moment, skewness, and the fourth moment, kurtosis of the standard normal distribution are zero.[4]

If a vector x is normally distributed, it follows that the z-score is standard normal. We can write this as

$$z \sim N(0, 1). \tag{1.5}$$

In mathematics, the tilde symbol $\sim$ means "approximately equal to", for example, $\pi \sim 22/7$. However, since statisticians have their pride and don't like that many mathematicians look down on them ☺ (for a complete list of reasons why mathematicians don't like statistics, see Appendix A.1), in statistics the symbol $\sim$ is redefined as "is distributed as." So Equation (1.5) reads: The z-score is standard normally distributed.

Conclusion: So is the transformation of a scalar or a vector with the z-score legitimate? Yes. The transformation of scalars allows the comparison of scalars on different scales as we saw in Example 1.6 and Table 1.1. The standardization of a vector is also fine in most cases and, as mentioned, often facilitates the analysis. Some concepts just work better with standardized data such as the principal component analysis (which finds the relative impact of regressors), regularization (which adds a bias to improve the overall regression result) or nearest neighbor search (an AI learning algorithm to classify data (e.g., mail versus spam mail)).

---

[4]To be precise, the *excess* kurtosis of the standard normal distribution is zero. The kurtosis of the standard normal distribution is 3. We then simply subtract 3 to get the excess kurtosis of 0. This is done so that the standard normal distribution can serve as a benchmark for the kurtosis of other distributions. For more info, see www.dersoft.com/normaldist.xlsx, cell Q16.

*Data Transformation to Standard Normal in the Copula
Correlation Model*

However, the transformation of non-normal data to standard normal
can cause problems. This was the case with the Copula correlation
model, which was blamed for the 2007–2009 global financial crisis.
I will put this one into Appendix A.2 since it is quite math-heavy.
A good read though for math nerds...

Transformation to Percentages

The transformation of absolute numbers to percentages sounds
absolutely harmless and it basically is. In fact, percentage changes
are typically more informative. Let's look at two examples:

Example 1.7: *Hedge funds A and B both own one stock. Hedge fund
A's stock increases from $1 to $2. Hedge fund B's stock increases
from $100 to $101. Hedge fund B argues that the performance is
equal since both stocks increased by $1. True? Yes and No. Yes,
both stocks increased by $1, however, the percentage changes are very
different.*

*We measure percentage changes of data with respect to time with
the equation*

$$\% change = \frac{S_{t_1} - S_{t_0}}{S_{t_0}}, \tag{1.6}$$

*where S_{t_x} is the stock price at time t_x. So for hedge fund A's stock
we have ($101 − $100)/$100 = 0.01 = 1% and for hedge fund B's
stock we have ($2 − $1)/$1 = 1 = 100%. So in percentage terms,
hedge fund B has performed much better! In conclusion, percentage
differences are typically more informative. Absolute differences can
be quite misleading.*

Standard Deviation versus Volatility

If you ever took a stats class, you will remember that standard
deviation measures the dispersion of the variable from its mean, i.e.,
how much a variable fluctuates. Volatility actually does the same

thing, just for percentage changes. So we can define:

> *Volatility is the Standard Deviation of Returns*

In finance, a return measures the performance of a money manager or simply a stock, defined as a percentage change as in Equation (1.6).

Why is working with percentage changes important? Let me explain with an example.

Example 1.8: *Hedge fund A owns stock X which has moved 1, 2, 3. Hedge fund B owns stock Y, which has moved 100, 101, 102. In finance, the fluctuation intensity, which can be measured either by standard deviation or volatility, is a measure of risk. The argument is, the higher the fluctuation, the more an asset can decline. True, however, high fluctuation also means an asset can increase a lot, so high fluctuation intensity also means high up-side potential. Skilled investors (like me ☺) typically like high-risk — high-potential stocks. You just have to have the right ones...*

Let's go back to measuring risk with fluctuation intensity. Asset X and Y have the same standard deviation of 1 (see Appendix A.3 for the calc). Therefore, hedge fund A argues, that the risk of asset X and Y are the same. Is this true or statistically misleading?

We can argue it is misleading because in percentage terms asset X fluctuates much more that asset Y. The volatility concept, working with relative changes, captures this percentage difference: Asset X has a volatility of 20.34%, asset Y a volatility of 0.0069%. And that is why all the smart finance people use the volatility concept and not standard deviation, which makes sense. For a derivation of standard deviation and volatility, see www.dersoft.com/StdvsVol.xlsx (case sensitive). If you are interested in the mathematical nitty-gritty, I put the equations and calculations into Appendix 1.3. That was a lot of tedious work, so give me some credit! ☺

Conclusion: Transforming absolute values to percentages has nothing to do with statistical cheating, but is absolutely legitimate. In fact, working with absolute changes can result in misleading information, whereas working with percentage changes is typically more informative and often gives us a clearer picture of the true characteristics of the data.

After discussing input data manipulation, let's now analyze timeframe snooping.

1.2. Manipulating the Research Timeframe

Many statistical research projects, especially in economics and finance, involve analyzing data in time. So some researchers manipulate the research time frame. Let's have a look how it's done.

1.2.1. *Timeframe Snooping*

> *Timeframe Snooping is selecting a time frame for which the research hypothesis works*

Associations often change in time. Figure 1.6 shows a nice example.

For the data in Figure 1.6, we derived the correlation coefficient[5] for different time frames. The correlation coefficient for the entire

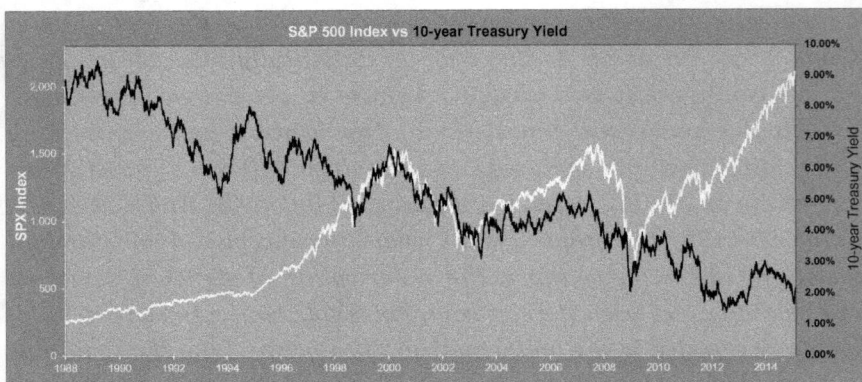

Figure 1.6: Different associations for different time frames for the S&P 500 (lighter function) and the 10-year Treasury Yield (darker function).

---

[5]A correlation coefficient measures the direction and strength of a linear relationship. See Chapter 2, Section 2.1 for details.

time-period from 1 January 1988, to 15 February 2015, is strongly negative at −0.770, since on average the S&P increased from 1988 to 2015 and the 10-year Treasury yield decreased. However, for sub-periods, the correlation coefficient is quite different: the correlation is negative (−0.429) from 1 January 1988, to 31 August 1998, positive from 1 September 1998, to 31 April 2009 (+0.600), and negative again from 1 May 2009, to 15 February 2015 (−0.397). The example of Figure 1.6 confirms that correlations can vary strongly with respect to the selected time period.

So a researcher who wants to show that there is a positive correlation between the S&P 500 and the 10-years Treasury yield, can simply use the time frame from 1 September 1998, to 31 April 2009, to derive the positive correlation, although, for the total time frame from 1 January 1988 to 15 February 2015, the correlation is negative.

1.2.2. *Lengthening and Shortening the Time Frame*

Researchers, keen to publish their papers, can also lengthen or shorten a time frame. Let's look at Figure 1.7.[6]

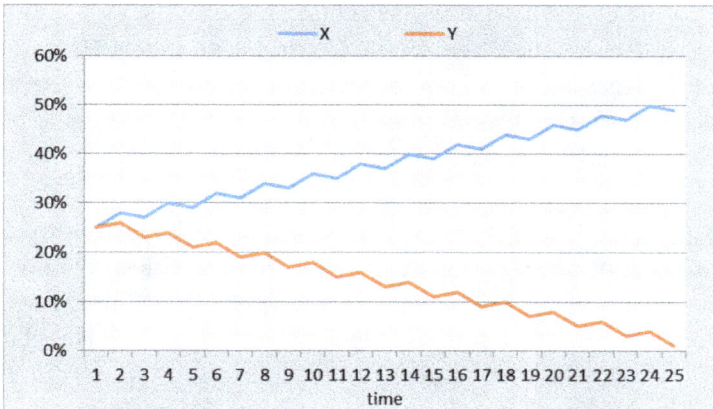

Figure 1.7: Two variables which are positively correlated short term, but negatively correlated long term.

---

[6]The idea for Figures 1.7 and 1.9 goes back to Paul Wilmott.

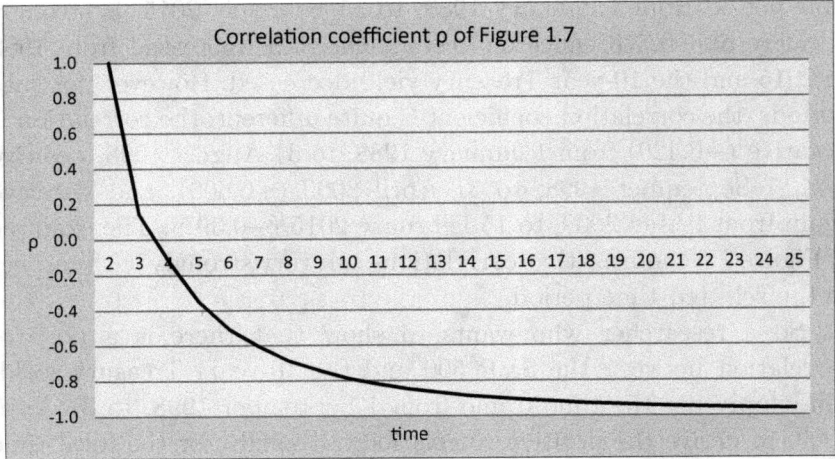

Figure 1.8: Correlation coefficient derived from Figure 1.7.

From Figure 1.7, we observe that the variables X and Y are positively correlated in the short term (both go up and down together), but negatively correlated in the long term, since X increases and Y decreases. This is confirmed when deriving the correlation coefficient in time, as seen in Figure 1.8.

Figure 1.8 verifies the higher correlation short term, which however decreases for longer timeframes. Hence a researcher can just use a short timeframe when he/she wants to derive a positive relationship, and a longer time frame when he/she wants to show a negative relationship!

We perform the opposite exercise of Figure 1.7 in Figure 1.9.

From Figure 1.9, we observe that the variables X and Y are negatively correlated in the short term (variable X increases when variable Y decreases and vice versa), but positively correlated in the long term, since X and Y both increase long term. This is confirmed when deriving the correlation coefficient in time, as seen in Figure 1.10.

Figure 1.10 verifies the lower correlation short term, which however increases for longer timeframes. Hence a researcher can just use a short timeframe when he/she wants to derive a negative

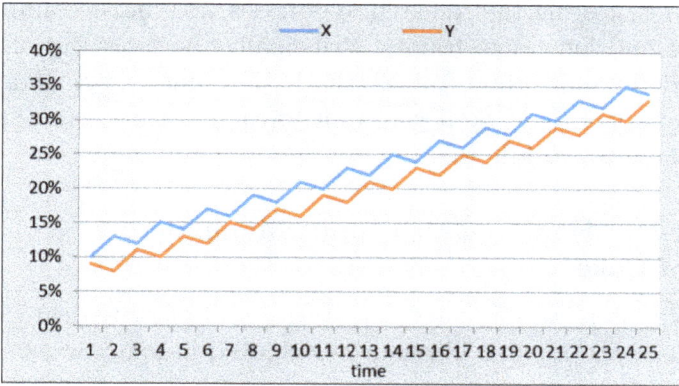

Figure 1.9: Two variables which are negatively correlated short term, but positively correlated long term.

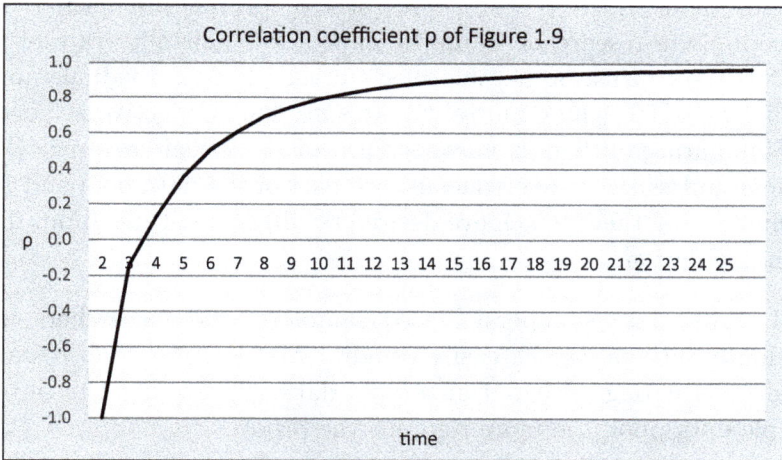

Figure 1.10: Correlation coefficient derived from Figure 1.9.

relationship and a longer time frame when he/she wants to show a positive relationship!

Conclusion: Associations often change in time. Therefore time frame snooping is a powerful tool of statistical cheaters! By selecting a specific time frame, often a desired result can be easily derived. How to catch time frame snoopers? That's pretty easy: Out-of-sample testing, i.e., testing the research hypothesis with out-of-sample data for short and long time frames. Principally the more out-of-sample data and the more time frames, the better, since more data means more statistical rigor. So if you want to catch a time frame snooper, be prepared to work hard!

Conclusion: How to Catch Input Manipulators — Piece of Cake

In this chapter, we discussed a powerful tool of statistical cheaters: Manipulate the inputs! This can be achieved in many ways:

1. Data Snooping, one of the most popular methods of cheating, is selecting specific data for which the research hypothesis works.
2. Biased sampling is another way of cheating and involves research with data which is not representative of the population. For example, a researcher wants to show a low mortality rate of the COVID19 virus. So he/she selects a sample of individuals under the age of 50, where mortality rates are low. Objective achieved!
3. Eliminating outliers is another form of statistical cheating. Outliers, unless they are erroneous, are part of the data pool and just eliminating them to achieve desired results, constitutes input data manipulation.

Luckily, data snoopers, biased samplers, and researchers who eliminate outliers can be easily caught: We just prove the research hypothesis wrong with out-of-sample data. The bigger the out-of-sample data pool, the more rigorous the proof.

4. A blunt way of statistical input data cheating is just using false input data. This was the method of Korean researcher Hwang Woo-suk, who claimed to have achieved cloning of human embryos. There was just one problem: His "research results" were human embryos from donor women.

In statistics, we often transform input data, for example transforming absolute data to percentages, or we standardize data so that the data distribution has a mean of 0 and a standard deviation of 1. This is statistically legitimate. In fact, percentage changes often provide more valuable information than absolute numbers, as seen in the superiority of the volatility concept compared to standard deviation. Standardizing data often leads to better results as in principal component analysis, regularization, or nearest neighbor search.

So, in conclusion, it is quite easy to catch data input manipulators, if we just put in the work and test their research with out-of-sample data and out-of-sample time frames. And blunt cheaters, who just use deliberately false input data, in the end often get caught. A satisfying realization...

Questions and Problems

The answers are available to instructors, please email gunter@dersoft.com

1. What is Data Snooping? Give an example of Data Snooping.
2. How can we catch Data Snoopers?
3. What is Biased Sampling? Give an example of Biased Sampling.
4. How can we catch Biased Samplers?
5. Is eliminating outliers statistical cheating?
6. Is data transformation of absolute data to percentage data statistical cheating?
7. What is standardization of data? Is standardization of data statistical cheating?
8. What is Time Frame Snooping? Give an example.
9. Associations often change in time. How do statistical cheaters exploit this property?
10. How can we catch Data Snoopers, Biased Samplers, Eliminating Outliers, and Time Frame Snoopers?

"The best thing about Statistics is that you get to play in everyone's backyard."

(John Turkey)

Appendix A.1: The Complete List Why Mathematicians Look Down on Statisticians

1. Statistics does not have a solid foundation with axioms and corollaries on which concepts and solution are based. "There is nothing to hold on to" as mathematicians complain.
2. Descriptive Statistics does what the name suggests: It describes variables, data and their distribution. This is not overly complex, "trivial" as mathematicians flex. And there are no proofs. What is a science without proofs?!
3. Statistics is probabilistic. Just as Albert Einstein hated Quantum Physics due to its uncertainty, "God does not play dice with the universe", mathematicians despise uncertainty. Science should have unique and exact results, not results which are only accepted at a confidence level of $x\%$!
4. Statistics has the concept of errors. A type I error is falsely rejecting the null hypothesis (e.g., the patient is healthy but diagnosed with an illness), a type II error is falsely accepting the null hypothesis (e.g., the patient is sick but diagnosed healthy). Science with an error concept? What on earth is that?
5. Statistics is just way too practical. Statisticians even take real-world data and analyze it! Even within mathematics, pure mathematicians often frown on applied mathematicians, who apply math to fields such as finance or AI to get a non-academic job! This superiority opinion of theory over application also exists in physics: When Sheldon asks Leonard if he is a theoretical or applied physicist and Leonard answers "applied", Sheldon replies with a disrespectful frown. Nerds know what I am talking about: The Big Bang Theory of course.
6. Statistics is just way too useful. It can analyze if a drug can cure an illness or a vaccine is efficient. Science has to be pure, elegant, and abstract. Beauty requires irrelevance!

Appendix A.2: Data Transformation in the Copula Model

> *"The most dangerous thing is when people believe everything that comes out of it [the Copula Model]."*
>
> (David Li)

In this section, we will show that changing non-normal data to standard normal can cause problems.

In 1959, the Japanese mathematician Abe Sklar formulated the Copula model. A simplistic version of the Copula model was transferred to finance by Oldrich Vasicek in 1987. In 2000, in a famous paper "On Default Correlation — A Ccopula model approach" David Li transferred the full model to finance.

The Copula model was first enthusiastically embraced, but fell into infamy during the Global Finance crisis in 2008 for which the Copula model was blamed. Numerous polemic papers were written, such as "Recipe for Disaster: The Formula that killed Wall Street", *Wired Magazine* (2009), "Wall Street Wizard (i.e., David Li) Forgot a Few Variables", *New York Times* (2009) or "The formula that felled Wall Street", *Financial Times* (2009). Let's have a closer look. The core equation of the Copula model is

$$C[Q_A(t), Q_B(t), \ldots, Q_n(t)]$$
$$= M_n[\underbrace{N^{-1}(Q_A(t))}_{\substack{\text{Transformation of} \\ Q_A(t) \text{ to standard normal}}}, N^{-1}(Q_B(t)), \ldots, N^{-1}(Q_n(t)); \rho_M], \quad (A.1)$$

where C is the Copula function, $Q_A(t)$ is the marginal distribution, which is a sub-distribution within the distribution M_n. In David Li's model, $Q_A(t)$ is the cumulative default probability of company A at time t, M_n is the n-dimensional normal distribution, N^{-1} is the inverse of the standard normal distribution, and p_M is the correlation structure of M.

Equation (A.1) reads: Given are the marginal distributions $Q_A(t)$ to $Q_n(t)$. There exists a Copula function C that allows the

Table A.1: Cumulative default distribution of company A and mapping to standard normal.

Year	Cumulative default probability of company A, $Q_A(t)$	Transformation of $Q_A(t)$ to standard normal via $N^{-1}(Q_A(t))$
1	6.51%	−1.5133
2	14.16%	−1.0732
3	21.03%	−0.8054
4	27.04%	−0.6116
5	32.31%	−0.4590
6	36.73%	−0.3390
7	40.97%	−0.2283
8	44.33%	−0.1426
9	47.17%	−0.0710
10	50.01%	0.0003

mapping of the marginal distributions $Q_A(t)$ to $Q_n(t)$ via N^{-1} to standard normal, and the joining of the (abscise values) $N^{-1}(Q_n(t))$ to a single, n-dimensional function $M_n[N^{-1}(Q_A(t)), N^{-1}(Q_B(t)), \ldots, N^{-1}(Q_n(t))]$ with correlation structure p_M.

Probably pretty heavy stuff for some. So let's look at an example:

Given are the cumulative default probabilities of company A in time (Table A.1).

Graphically the mapping of $Q_A(t)$ to standard normal is displayed in Figure A.1.

The mapping in Figure A.1 is done in two steps. First, a percentile-to-percentile transformation from the original $Q_A(t)$ distribution to the standard normal distribution is performed (up-arrows). In a second step, the abscise value (x-axis value) of the standard normal distribution is found (down-arrows). These abscise values $N^{-1}(Q_x(t))$ are then input into Equation (A.1).

To evaluate if the transformation from the non-normal default distribution to a standard normal distribution or another Copula flaw contributed to the global financial crisis 2008, we have to ask two questions:

(1) Did the Copula model contribute to the 2008 crisis?
(2) Was the transformation or any other Copula flaw a reason for the 2008 crisis?

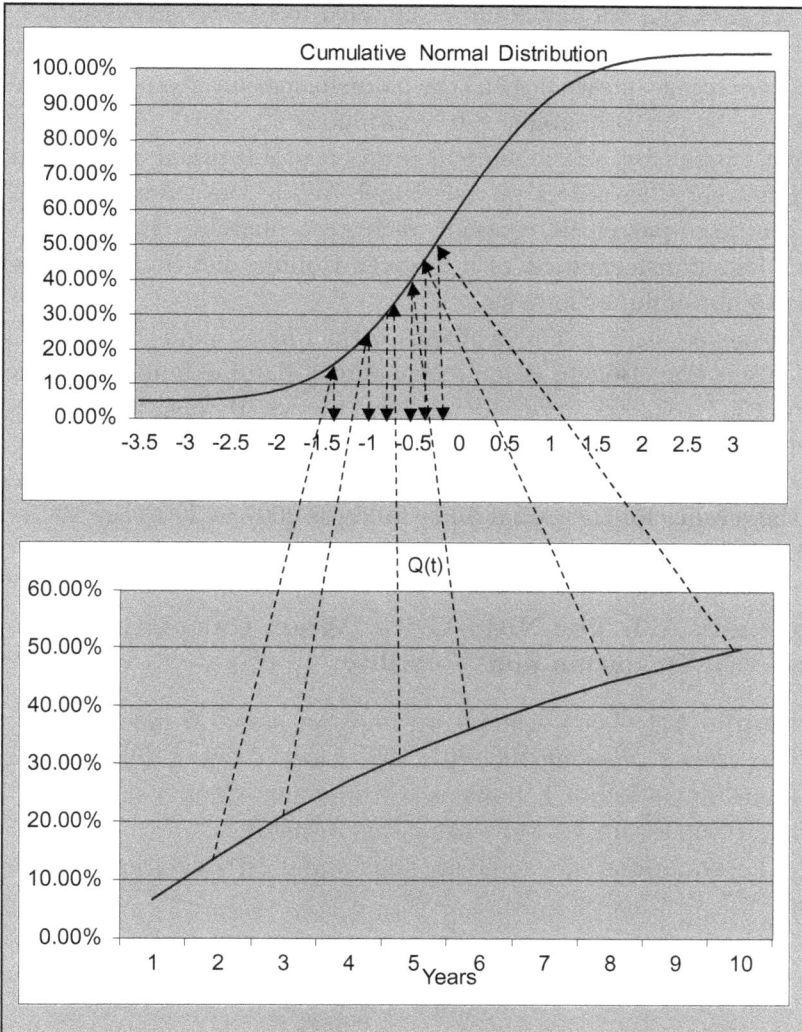

Figure A.1: Mapping of the non-normal default probabilities $Q_A(t)$ to a standard normal distribution.

Question (1) can be answered with yes. The Copula function was viewed as a Deus ex Machine, a solution to all problems of humanity, in particular, the problem of evaluating the default correlations between 125 assets in a collateralized debt obligation (CDO) in time.

Hence the CDO trading volume exploded from US\$64 billion in 2003 to US\$455 billion in 2006, forming the subprime mortgage bubble.

Regarding question (2), the transformation from non-normal discrete default probabilities to continuous standard-normal default probabilities for mathematical and computational convenience, is definitely a weakness of the model. While the transformation is percentile-to-percentile, specific default probability information is lost. The transformation of a discrete Copula distribution can also lead to non-unique Copulas.

However, the main problem with the Copula function was miscalibration. Benign default probabilities and default correlations from 2002 to 2006 were fed into the model. If false input data is input, it cannot be expected that a model works. *Garbage in — Garbage out* in programming terminology. For a detailed discussion on the reasons for the global financial crisis 2007 to 2009, see Meissner (2019, Chapter 6).

Appendix A.3: The Nitty-Gritty When Calculating Standard Deviation and Volatility

Example A.1: Let's assume we have an asset X moving 1, 2, 3, and an asset Y, moving 100, 101, 102. Asset X and Y have the same standard deviation of 1, however, a different volatility of 20.34% for X and 0.0069% for Y. Let's show this in detail:

The equations we need are the standard deviation Std of a stock S:

$$\text{Std(S)} = \sqrt{\frac{1}{n-1} \sum_{t=1}^{n} (S_t - \bar{S})^2}, \qquad (A.2)$$

where $\bar{S}$ is the arithmetic mean, and n is the number of data points

$$\bar{S} = \frac{1}{n} \sum_{t=1}^{n} S_t. \qquad (A.3)$$

Volatility takes percentage changes as inputs. Since a percentage change $\frac{S_{t_1} - S_{t_0}}{S_{t_0}} \sim \ln(S_{t_1}/S_{t_0})$, to calculate volatility, Equation (A.2) changes to

$$\text{Vol}(S) = \sqrt{\frac{1}{n-1} \sum_{t=1}^{n} (\ln(S_t/S_{t-1}) - \bar{S}_V)^2}, \qquad (A.4)$$

where ln stands for natural logarithm and

$$\bar{S}_V = \frac{1}{n} \sum_{t=1}^{n} \ln(S_i/S_{i-1}). \qquad (A.5)$$

Let's apply these equations to our example. Following Equation (A.3), the mean of X is

$$\bar{S}(X) = \frac{1}{n} \sum_{t=1}^{n} S_t = \frac{1}{3} \sum_{t=1}^{n} (1+2+3) = 2.$$

Following Equation (A.2), the standard deviation of X is

$$\text{Std}(X) = \sqrt{\frac{1}{n-1} \sum_{t=1}^{n} (S_t - \bar{S})^2}$$

$$= \sqrt{\frac{1}{2} \sum_{t=1}^{n} (1-2)^2 + (2-2)^2 + (3-2)^2 = 1}.$$

For asset Y we have

$$\bar{S}(Y) = \frac{1}{n} \sum_{t=1}^{n} S_t = \frac{1}{3} \sum_{t=1}^{n} (100 + 101 + 102) = 101$$

and from Equation (A.2), the standard deviation of Y is

$$\text{Std}(Y) = \sqrt{\frac{1}{n-1} \sum_{t=1}^{n} (S_t - \bar{S})^2}$$

$$= \sqrt{\frac{1}{2} \sum_{t=1}^{n} (100 - 101)^2 + (101 - 101)^2 + (102 - 101)^2 = 1}.$$

So the standard deviation concept is telling us that assets X and Y have the same dispersion from their mean, i.e., fluctuate the same. However, this is somewhat misleading since, in percentage terms, X fluctuates much more than Y. The volatility concept reveals this information:

We first have to calculate the volatility mean of X from Equation (A.5):

$$\bar{S}_V(X) = \frac{1}{n} \sum_{t=1}^{n} \ln(S_t/S_{t-1}) = \frac{1}{2} \sum_{t=1}^{n} \left[\ln\left(\frac{2}{1}\right) + \ln\left(\frac{3}{2}\right) \right] = 54.93\%.$$

Note that the number of inputs has decreased to $n - 1$, so in our example $3 - 1 = 2$, when calculating volatility.

The volatility of X, following Equation (A.4) is

$$\text{Vol}(X) = \sqrt{\frac{1}{n-1} \sum_{t=1}^{n} [\ln(S_t/S_{t-1}) - \bar{S}_V]^2}$$

$$= \sqrt{\frac{1}{1} \sum_{t=1}^{n} [\ln(2/1) - 0.5493]^2 + [\ln(3/2) - 0.5493]^2} = 20.34\%.$$

For asset Y the volatility mean from Equation (A.5) is

$$\bar{S}_V(\dot{Y}) = \frac{1}{n} \sum_{i=1}^{n} \ln(S_t/S_{t-1})$$

$$= \frac{1}{2} \sum_{t=1}^{n} \left[\ln\left(\frac{101}{100}\right) + \ln\left(\frac{102}{101}\right) \right] = 0.99\%.$$

Following Equation (A.4), the volatility of asset Y is

$$\text{Vol}(Y) = \sqrt{\frac{1}{n-1} \sum_{t=1}^{n} [\ln(S_t/S_{t-1}) - \bar{S}_V]^2}$$

$$= \sqrt{\frac{1}{1} \sum_{t=1}^{n} [\ln(101/100) - 0.0099]^2 + [\ln(102/101) - 0.0099]^2}$$

$$= 0.0069\%.$$

Phew, that was tedious work! It may be easier to just look at the spreadsheet www.dersoft.com/StdvsVol.xlsx (case sensitive). Nevertheless, we find:

Conclusion: The standard deviation concept tells us that assets X and Y have the same dispersion from the mean, i.e., fluctuate the same. This is quite misleading since, in percentage terms, X and Y have very different fluctuations. The volatility concept reveals these percentage changes. Therefore it can be argued that volatility is a superior concept to standard deviation, at least for time series data. For a simple spreadsheet, calculating standard deviation and volatility, see www.dersoft.com/StdvsVol.xlsx.

Every mathematical software package such as Excel, SAS, R or Python has a built-in equation for the standard deviation. Let's show how the Python code looks for our Example A.1.

```
import numpy as np

X=[1,2,3]

Y=[100,101,102]

print(np.std(X,ddof=1))   #ddof=1 guarantees that the sample standard
                           deviation with 1/(n-1) is applied

print(np.std(Y,ddof=1))


X1=[np.log(2/1),np.log(3/2)]   #this derives the natural log inputs for the
                               Volatility calculation

print(np.std(X1,ddof=1))


Y1=[np.log(101/100),np.log(102/101)]

print(np.std(Y1,ddof=1))

OUTPUT

1.0 #is the standard deviation of (1,2,3)
```

```
1.0 #is the standard deviation of (100,101,102)

0.20342194425645393   #is the volatility of (1,2,3)

6.932079621125092e-05   #is the volatility of (100,101,102)


Process finished with exit code 0
```

These results are naturally identical with the results in Example A.1.

The Python code can be run at https://repl.it/@GunterMeiss ner/StdvsVol#main.py.

References

Li, D., On default correlation — A Copula function approach. *Journal of Fixed Income*, Spring 2000, 9(4): 43–54.

Meissner, G., *Correlation Risk Modeling and Management*, 2nd edition. London: RISKbooks, 2019.

Chapter 2

Manipulating Statistical Calculations

"The death of one man is a tragedy. The death of millions is a statistic."

(Josef Stalin)

In this chapter, we will investigate how statistical cheaters manipulate the statistical calculations to derive desired results. We will discuss the two main areas of statistics, Descriptive Statistics, which finds the association between variables, and Inferential Statistics, which draws conclusions from Descriptive Statistics, such as determining if the associations happen by chance or are significant. Naturally, we have to go a bit into the statistical math. But no worries, it will be a walk in the park ☺.

2.1. Regression Analysis

Within descriptive statistics, regression analysis is a key field. We can define it as

> *Regression analysis finds the association between two or more variables*

We already looked at many examples of potential associations between variables in Chapter 1, such as vaccination and Autism, smoking and cancer, or SAT scores between public and private schools. More associations of interest, are

- Gun ownership and shootings in a country;
- CO_2 emissions and global warming; or
- Internet surfing and decreased creativity.

On the lighter side, we can investigate if there is an association between

- Pandemic boredom and GameStop trading;
- Marriage and longevity (married people actually do live longer, who would have thought ☺); or
- Money and Happiness (most studies show some evidence of a positive association, but there is decreasing marginal utility: The more you make, each additional dollar gives you less happiness than the previous one. These economics lectures are really paying off! ☺).

A key property of regression analysis is correlation. What is correlation? Well, correlation measures how two or more variables are associated, meaning how they relate to each other or move together in time. There are tons of correlation concepts. Figure 2.1 gives an overview, more than you probably ever wanted to know about correlation.

> *"No subject in this universe is unworthy of study"*
>
> (Karl Pearson)

In this book, we will focus on the most applied correlation model, the Pearson correlation model, which defines

> *Correlation measures the linear direction and strength of the association between of two or more variables*

Karl Pearson (1857–1936), was actually a universal genius. Not only did he bless us with inventing statistics, he was also a lawyer, Germanist, and physicist. He worked on the fourth dimension, time, and claimed that for an observer traveling at the speed of light, time would stand still, i.e., the observer would see "eternal now". And if traveling at speeds faster than light, we would have time reversal.

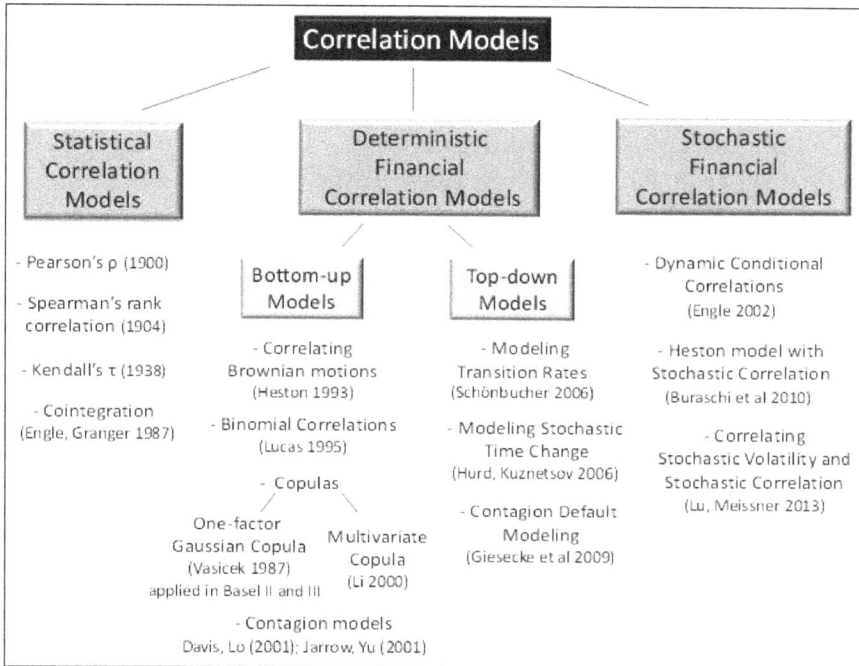

Figure 2.1: Correlation concepts.
Source: Meissner (2019).

This is all amazing and true, however, Albert Einstein stepped in and showed that the speed of light is the speed limit in the universe. So unfortunately there is no possibility of going back in time. Even traveling at the speed of light is almost impossible, since a subject trying to do so, would get infinitely heavy, and there is no infinite energy to propel a subject close to infinite weight further. Ok, enough physics, let's get back to cheating. . .

A Quick Recap of the Pearson Correlation Model

If we want to show how evil forces cheat with the Pearson model, we have to understand it. So let me walk you through the key points.

The Pearson correlation model measures the linear strength and direction of two variables with two critical parameters: (1) the correlation coefficient ρ; and (2) the regression coefficient β.

(1) The Correlation Coefficient ρ

Figure 2.2 gives an intuitive overview of the correlation coefficient ρ:

From the upper row in Figure 2.2, we observe that $\rho = 1$ or $\rho = -1$, if all data points lie on a straight line, which we call the regression function. The more the data set disperses, the lower the correlation coefficient. If the data set is distributed randomly in a circle, there is no correlation between the variables (middle data set in the upper row) and $\rho = 0$.

The middle row shows that the slope is irrelevant in the derivation of the correlation coefficient ρ: As long as all data points lie on an upward sloping line, $\rho = 1$, and as long as all data points lie on a downward sloping line, $\rho = -1$. However, the slope is critical and will be measured with the regression coefficient β, which we will discuss shortly.

From the third row of Figure 2.2, we see different data sets all resulting in $\rho = 0$. The parabola in the middle may seem confusing. We have a dependency $Y = X^2$, but the correlation coefficient is 0! This shows a drawback of the linear Pearson model. For certain

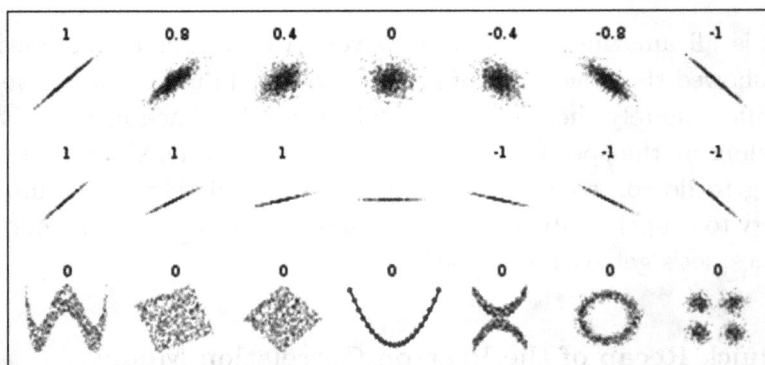

Figure 2.2: Pearson correlation coefficient ρ for different data sets.
Source: Denis Boigelot: https://www.upenn.edu/learninganalytics/MOOT/slides/W002V004.pdf.

nonlinear dependencies, the Pearson correlation often results in a low correlation, for a parabola it is 0. We will show in Appendix A.1 that independence implies uncorrelatedness (i.e., a correlation coefficient of 0), but uncorrelatedness does not imply independence, as we just realized with the parabola example.

The correlation coefficient is not defined for the horizontal function in the middle of the second row since the standard deviation of Y is zero, which is in the denominator of the correlation coefficient calculation, see Equation (2.4).

To calculate the correlation coefficient between two variables, we need some basic equations.

The arithmetic mean of a number series S is

$$\bar{S} = \frac{1}{n} \sum_{t=1}^{n} S_t. \tag{2.1}$$

The standard deviation, which measures the dispersion from the mean is

$$\sigma_S = \sqrt{\frac{1}{n-1} \sum_{t=1}^{n} (S_t - \bar{S})^2}. \tag{2.2}$$

The covariance measures how two variables "co-vary" together. If two variables X and Y move up together and down together, the covariance will be highly positive. If the two variables move in opposite direction, i.e., X moves up when Y moves down and vice versa, the covariance will be negative. The equation for the covariance is

$$\mathrm{COV}_{XY} = \frac{1}{n-1} \sum_{t=1}^{n} (x_t - \bar{x})(y_t - \bar{y}). \tag{2.3}$$

The covariance takes values between $-\infty$ and $+\infty$. The correlation coefficient ρ is a conveniently scaled covariance which takes values between -1 and $+1$.

$$\rho_{XY} = \frac{\mathrm{COV}_{XY}}{\sigma_X \, \sigma_Y}, \tag{2.4}$$

where σ_X and σ_Y are the standard deviations of X and Y, as defined in Equation (2.2)

Table 2.1: Prices of assets X and Y and their returns.

Year	Asset X	Asset Y	Returns x (%)	Returns y (%)
2015	$100	$200		
2016	$150	$270	40.55	30.01
2017	$125	$460	−18.23	53.28
2018	$150	$410	18.23	−11.51
2019	$160	$480	6.45	15.76
2020	$280	$380	55.96	−23.36
		Average =	**20.59**	**12.84**

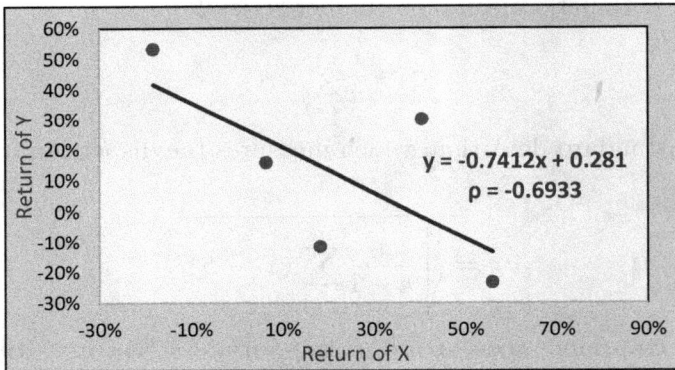

Figure 2.3: Returns of assets X and Y, the regression function $y = -0.7412x + 0.281$ and the correlation coefficient $\rho = -0.6933$.

Why don't we calculate the correlation coefficient ρ_{XY} with a numerical example? Let's assume we have asset prices as in Table 2.1.

The returns in Table 2.1 are derived as $\ln(S_t/S_{t-1})$, where ln is the natural logarithm. We could also have used percentage changes $(S_t - S_{t-1})/S_{t-1}$ to calculate the return. The reason why we typically work with logarithmic changes is that they are additive in time, whereas percentage changes are not. This is actually a potential cheating method, see Chapter 3, Section 3.1 for details.

Figure 2.3 shows the returns of assets X and Y per year, the correlation coefficient $\rho = -0.6933$, and the regression coefficient $\beta = -0.7412$ which we will discuss shortly.

OK, here come the calculations for the correlation coefficient ρ. It may be easier to just look at the Excel spreadsheet www.dersoft .com/Pearsonmodel.xlsx, cells P19 and P21. However, here are the calculations:

Following Equation (2.2), the standard deviation of the returns of asset X is

$$\sigma_X = \sqrt{\frac{1}{n-1} \sum_{t=1}^{n} (x_t - \overline{x})^2} = 0.2899.$$

For the detailed calculation of σ_X, see Appendix A.1 and the spreadsheet www.dersoft.com/Pearsonmodel.xlsx cell P14.

For the standard deviation of the returns of asset Y, we have

$$\sigma_Y = \sqrt{\frac{1}{n-1} \sum_{t=1}^{n} (y_t - \overline{y})^2} = 0.3099.$$

For the calculation of σ_Y see Appendix A.1 and the spreadsheet ww w.dersoft.com/Pearsonmodel.xlsx cell T14.

Following Equation (2.3), the covariance of the returns of assets X and Y is

$$\text{COV}_{XY} = \frac{1}{n-1} \sum_{t=1}^{n} (x_t - \overline{x})(y_t - \overline{y}) = -0.0623.$$

for the math, see Appendix A.1 or www.dersoft.com/Pearsonmodel .xlsx cell L14.

OK, we are ready to derive the correlation coefficient ρ from Equation (2.4):

$$\rho_{XY} = \frac{\text{COV}_{XY}}{\sigma_X\,\sigma_Y} = \frac{-0.0623}{0.2899x0.3099} = -0.6933.$$

Interpretation: Since the correlation coefficient is defined $-1 \leq \rho \leq +1$, we can conclude that the returns of assets X and Y are strongly negatively correlated. We kind of knew that already by looking at Figure 2.3, but now we have statistical evidence. Fantastic!

(2) The Regression Coefficient β

Since, as already stated above, the slope of the regression function is irrelevant in the derivation of the correlation coefficient ρ, we have to calculate the second important regression parameter, the regression coefficient β, which actually *is* the slope of the regression function.

The equation is nice and easy:

$$\beta = \frac{\text{COV}_{XY}}{\sigma_X^2}. \tag{2.5}$$

A proof of Equation (2.5), which I am sure the reader can't wait to see, is at www.dersoft.com/Regressionproof.pptx.

So for our example we have

$$\beta = \frac{-0.0623}{(0.2899)^2} = -0.7412.$$

Interpretation: The slope of the regression function is -0.7412. So on average, for every increase in the return of asset X of 1%, the return of asset Y decreases by -0.7412%.

The Python code using the numpy library to derive the correlation coefficient ρ and the regression coefficient β is straightforward:

```
import numpy as np

X=[0.4055,-0.1823,0.1823,0.0645,0.5596]
Y=[0.3001,0.5328,-0.1151,0.1576,-0.2336]

print(np.corrcoef(X,Y))

CovXY=(np.cov(X,Y,ddof=1))   #ddof=1 guarantees that the sample covariance with 1/(n-1) is applied
VarX=(np.var(X,ddof=1))
print(CovXY/(VarX))   # derives the regression coefficient β from equation (2.5)
```

```
OUTPUT (which comes in matrix form)

[[ 1.          -0.69320535]        ⌉  ρ(X,X)      ρ(X,Y)
 [-0.69320535  1.         ]]        ⌋  ρ(Y,X)      ρ(Y,Y)

[[ 1.          -0.74105555]        ⌉  β(X,X)      β(Y,X)
 [-0.74105555  1.14281969]]        ⌋  β(Y,X)      cov(Y,Y)/Var(X)

Process finished with exit code 0
```

The code can be run at https://repl.it/@GunterMeissner/Rho-and-Beta#main.py.

The results are identical (who would have thought ☺) with the results we derived from scratch in Appendix A.1. See www.dersoft.com/Pearsonmodel.xlsx, cells P23 and P25 for the derivation in Excel.

Before we discuss cheating opportunities with ρ and β, let's see if they are related. Indeed, they are by the equation

$$\beta = \rho \frac{\sigma_Y}{\sigma_X}. \tag{2.6}$$

Equation (2.6) makes sense intuitively, I hope: The higher the standard deviation of Y, σ_Y, the more dispersed are the Y-values, in simple words "stretched out" on the Y-axis, leading to a higher slope β of the regression function. Vice versa, the higher the standard deviation of X, σ_X, the more dispersed are the X-values, so the more "stretched out" on the X-axis, leading to a lower slope of the regression function.

So how can we cheat with ρ and β? Easy. Sometimes we have a situation where only ONE of the two critical parameters is significant. For example, if the standard deviation of Y, σ_Y, is much higher than the standard deviation of X, σ_X, it follows from Equation (2.6) that β is much higher than ρ. Hence "creative reporting" would just be reporting β and not ρ to prove significance. For $\sigma_X \gg \sigma_Y$, the opposite applies. We will discuss an example of false reporting with respect β and ρ in Chapter 3, Example 3.2.

2.1.1. *Applying the Linear Pearson Model to Nonlinear Data*

One of the most significant limitations of the Pearson correlation model is that it measures the *linear* strength of a relationship. However, most phenomena in natural sciences such as physics, biology or astronomy as well as social sciences such as economics and finance are nonlinear. Let's look at some examples, in which the Pearson models gives erroneous results.

We already derived from Figure 2.2 that a parabola $Y = X^2$ results misleadingly in a correlation coefficient of zero. Let's have a closer look.

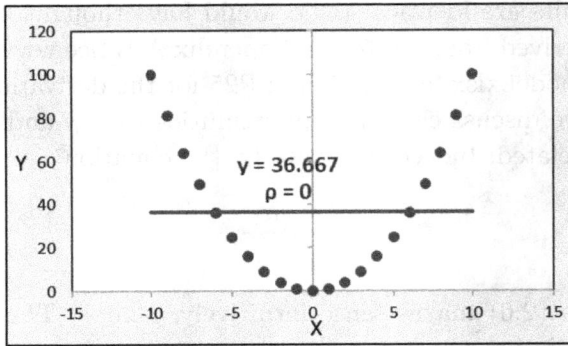

Figure 2.4: Fitting the parabola $Y = X^2$ with the linear Pearson model.

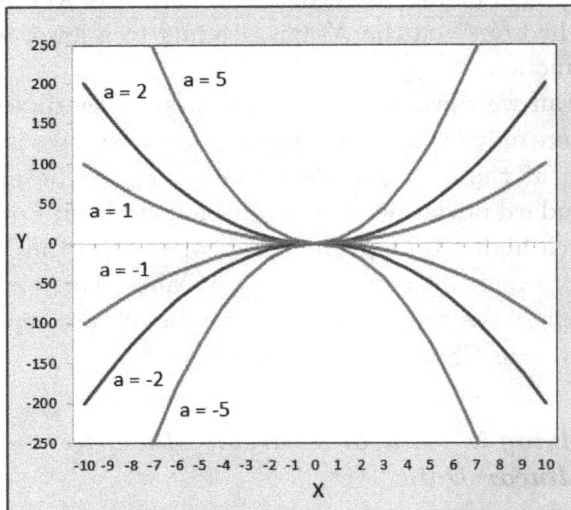

Figure 2.5: Parabolas $Y = aX^2$, for different values of "a", all resulting in $\rho = 0$.

As we can see from Figure 2.4, the fit is horrible with a horizontal regression function $Y = 36.667$ and correlation coefficient of $\rho = 0$. In fact, all parabolas with $Y = a\,x^2$, $a \neq 0$, result in a correlation coefficient of 0, as displayed in Figure 2.5.

Figures 2.4 and 2.5 can be found at www.dersoft.com/underfitt ing.xlsx.

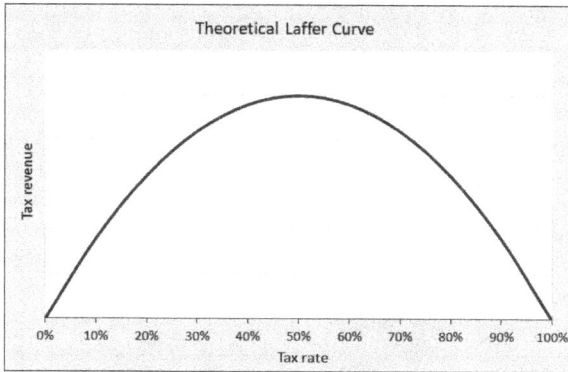

Figure 2.6: Theoretical Laffer curve: Tax revenue as a function of the tax rate.

Hence, we conclude that the Pearson model cannot be applied for certain nonlinear functional relationships. The relationship between variables has to be approximately linear for the Pearson model to produce sensible results. And of course this enables forms of cheating:

Two Practical Examples of Potential Misuse of the Pearson Model

During his presidency from 1981 to 1989, Ronald Reagan cut the marginal income tax rate in the US from 70% to a remarkable 28%! He justified the tax cut with the claim that economic activity would increase, and tax revenue would therefore, despite the tax cut, also increase. Sounds all too good to be true, and it was.

Underlying the claim was the Laffer curve, a theoretical relationship between tax revenue and the tax rate, popularized by supply-side economist Arthur Laffer. The theoretical Laffer curve is displayed in Figure 2.6.

The theoretical Laffer curve in Figure 2.6 does make sense: For a tax rate of 0%, there is no tax revenue. For a theoretical tax rate of 100%, citizens would emigrate or evade taxes, so the tax revenue would again be 0. The maximum tax revenue is at a 50% tax rate since Figure 2.6 is symmetrical.

The correlation coefficient ρ of the theoretical Laffer curve is zero (see a simple spreadsheet www.dersoft.com/Laffercurve.xlsm,

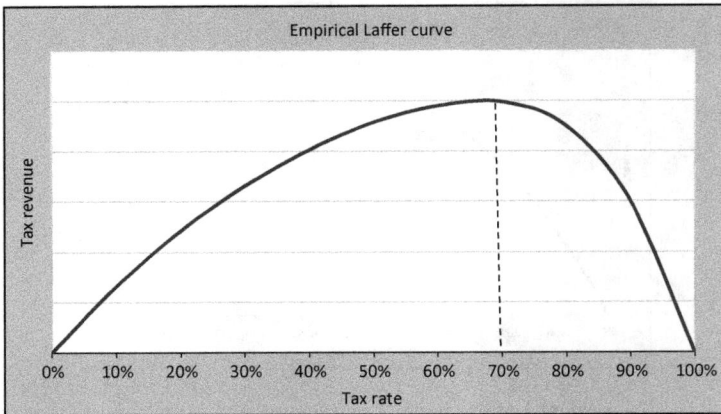

Figure 2.7: Empirical Laffer curve with a maximum tax revenue at a tax rate of about 70%.

cell N20) since the Pearson model is not able to properly assess the nonlinear relationship in Figure 2.6. So a researcher could argue that there is no correlation between the tax rate and tax revenue, which is statistically correct if applying the Pearson correlation model, however extremely misleading, since there is obviously a dependency of tax revenue with respect to the tax rate.

Numerous empirical studies on the Laffer curve exist, see for example Goolsbee 2010, Trabandt and Uhlig 2011, or Lundberg 2017. It turns out that the empirical relationship between tax revenue and the tax rate is as displayed in Figure 2.7.

The empirical Laffer curve shows why Reagan's tax cuts did not work: A decrease in the marginal tax rate from 70% to 28% decreases the tax revenue significantly, not increases it!

In the past, other politicians have tried and failed with the same strategy. In 2012 and 2013, the governor of Kansas, Sam Brownback cut the marginal income tax rate by almost 30%, expecting "a shot of adrenaline into the heart of the Kansas economy" and anticipated that the tax cuts were tax revenue neutral. However, following Figure 2.7, tax revenues plunged, leading to cuts in education and other vital services. In 2017, the Kansas legislature reversed Brownback's tax cuts. So whenever you find a politician claiming

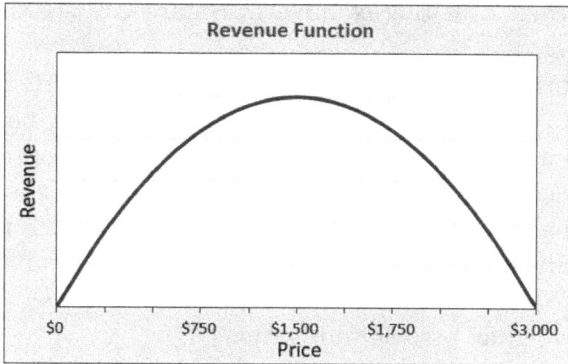

Figure 2.8: Revenue as a function of price.

that tax cuts will stimulate the economy and result in unchanged or even increased tax revenue, send him/her a copy of this book ☺.

The correlation coefficient ρ of Figure 2.7 is 0.2790%, (see spreadsheet www.dersoft.com/Laffercurve.xlsm, cell T20) because most of the relationship between the tax rate and tax revenue is positive. This shows that the Pearson model, for certain nonlinear relationships, gives at least a sensible approximation of the correlation between the nonlinear data.

In social sciences, we find "mountain shaped" relationships as in Figures 2.6 and 2.7 quite often. For example, the Revenue function R in economics is mountain shaped. The equation is simply

$$R = px \tag{2.7}$$

where R is the revenue, p is the price of a unit sold, and x is the sold units.

An example of Equation (2.7) would be APPL selling $x = 1,000,000$ iPhones for $p = \$1,000$ each. So the Revenue would be $R = px = 1,000,000 \times \$1,000 = \$1,000,000,000$. Graphically, the Revenue function is displayed in Figure 2.8.

An example of the revenue function in Figure 2.8 may be iPhones: At a theoretical price of 0, there would certainly be high demand, but no revenue. And at a price of \$3,000 there would be, at least in the example of Figure 2.8, no demand. The price per iPhone which maximizes revenue in the example of Figure 2.8 is \$1,500. Sounds about right...

Importantly, the correlation coefficient ρ in Figure 2.8 is zero (see www.dersoft.com/Revenue.xlsm, cell N20), implying no correlation between revenue and price. This again shows the limitation of the Pearson model with respect to nonlinear data since there is clearly an association between revenue and price. A cheater may exploit the limitation of the Pearson model, which typically underestimates the correlation between nonlinear variables, claiming little or no association when an association exists.

How can we catch cheaters who try to mitigate the association between variables by using the Pearson model? Easy. We need to apply nonlinear regression methods, which is exactly what we are going to do in the next chapter.

2.1.2. *Polynomial Regression: Under- and Overfitting*

Nonlinear data can typically be better fitted with nonlinear regression methods. The most popular method is polynomial fitting. A polynomial is a simple mathematical expression that involves only the operations of addition, subtraction, multiplication, and non-negative integer exponents. So $x^2 - 4x + 5$ is a polynomial, but X^{-2} is not.

We can define Underfitting as

> *Underfitting is using a regression function which is not complex enough to capture the relationship between variables*

Underfitting typically involves fitting a too low degree polynomial to the data. And we did that already in Figure 2.4, where we fitted a linear regression, which is a polynomial of degree 1, through the second-order function $Y = X^2$. So we are already done with underfitting! Excellent!

So let's look at Overfitting. We can define it as

> *Overfitting is using a too complex regression function which does not represent the general relationship between variables well*

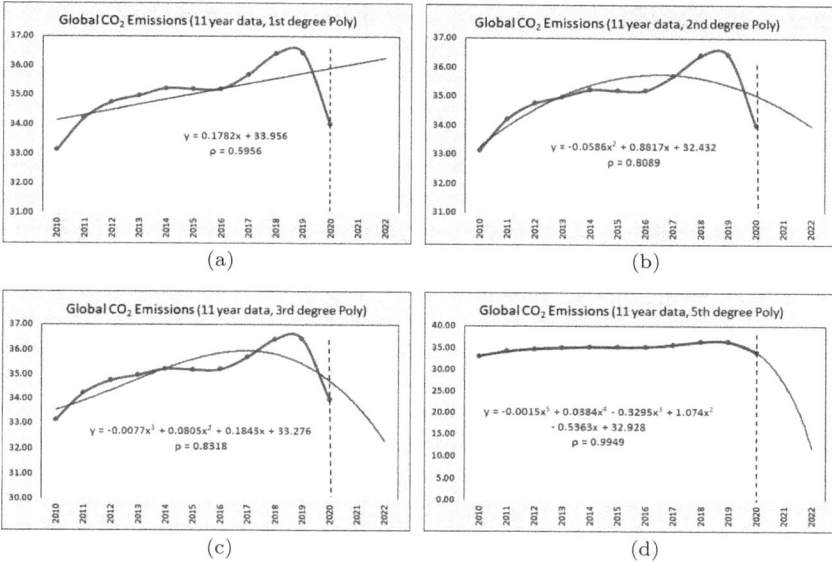

Figure 2.9: Global CO$_2$ emissions (function with markers), different degree polynomials (function without markers) from 2010 to 2020. The data to the right of the vertical dashed line is the forecast, derived by extrapolation of the polynomial regression function.

Overfitting typically leads to bad forecasts since the data in the training set cannot be generalized to out-of-sample data. Overfitting means fitting a too high degree polynomial through the data. The data is often fit too closely to the data points, not representing the general relationship between the two variables well. Let's look at a real-world example. Figure 2.9 shows the CO$_2$ emissions from 2010 to 2020, so for the last 11 years.

What are the takeaways from Figure 2.9? We observe low CO$_2$ emissions in 2020 because of the Covid19 economic slowdown. A linear regression function in Figure 2.9(a) represents the overall increasing CO$_2$ emissions, and forecasts an increase beyond 2020 despite the outlier in 2020. However, the higher degree polynomials in Figures 2.9(b)–2.9(d) are strongly impacted by the 2020 outlier and forecast lower CO$_2$ emissions beyond 2020.

So what is the correct forecast? We do observe the better fit of the data for higher degree polynomials, since the higher the polynomial degree, the higher the correlation coefficient ρ, which is displayed in

Figures 2.9(a)–2.9(d). The question is whether Figures 2.9(b)–2.9(d) are overfitting the data and leading to a false forecast! To answer this question, we have to look at the probability of recurrence of the outlier: Pandemics are rare events. The last pandemic with a high social and economic impact was the Spanish flu from 1918 to 1920, which killed approximately 30 million people. We encountered SARS in 2002/2003 with 811 mortalities, all outside the US, and the Swine flu pandemic in 2009 with about 284,000 tragic deaths. However, the economic impact of the Swine flu was low.

So arguably a rare outlier such as the Covid19 pandemic, whose reoccurrence probability is low, should not impact a forecast significantly. However, this is exactly the case in Figures 2.9(b)–2.9(d): High degree polynomials weigh more recent data high to derive a forecast. In addition, we only have 11 data points in Figures 2.9(b)–2.9(d). Therefore the relative weight of an outlier is high. To derive a more rigorous CO_2 forecast, we should use a bigger data set, which is exactly what we are doing in Figure 2.10.

The main take-away from Figure 2.10 when compared to Figure 2.9 is that the outlier in 2020 is less significant. Figures 2.10(a)–2.10(c) with 1st, 2nd, and 3rd degree polynomials forecast an increase in CO_2 emissions in contrast to Figures 2.9(b)–2.9(d). The 5th degree polynomial in Figure 2.10(d) however weighs recent data strongly and forecasts a decrease. Figures 2.9 and 2.10 (and more polynomials) can be found at www.dersoft.com/polynomials.xlsx.

The Python code to derive a polynomial fit is straight forward. Let's derive the 5th degree polynomial for CO_2 emissions from 2010 to 2020, as displayed in Figure 2.9(d):

```
import numpy as np
t = [1,2,3,4,5,6,7,8,9,10,11]
Y = [33.13, 34.22, 34.76, 34.98, 35.23, 35.20, 35.21, 35.70,
36.41,36.44, 34.00]

degree = 5
print(np.polyfit(t,Y,degree))

OUTPUT
[-1.53205128e-03  3.84090909e-02 -3.29495921e-01  1.07398019e+00
 -5.36279720e-01  3.29284848e+01]
```

The first term of the output is to be raised to the fifth power, the second term to the fourth power, and so on. The result is identical with the Excel result in Figure 2.9(d), who would have thought ☺. The Python code can be run at https://repl.it/@GunterMeissner/Polynomial#main.py.

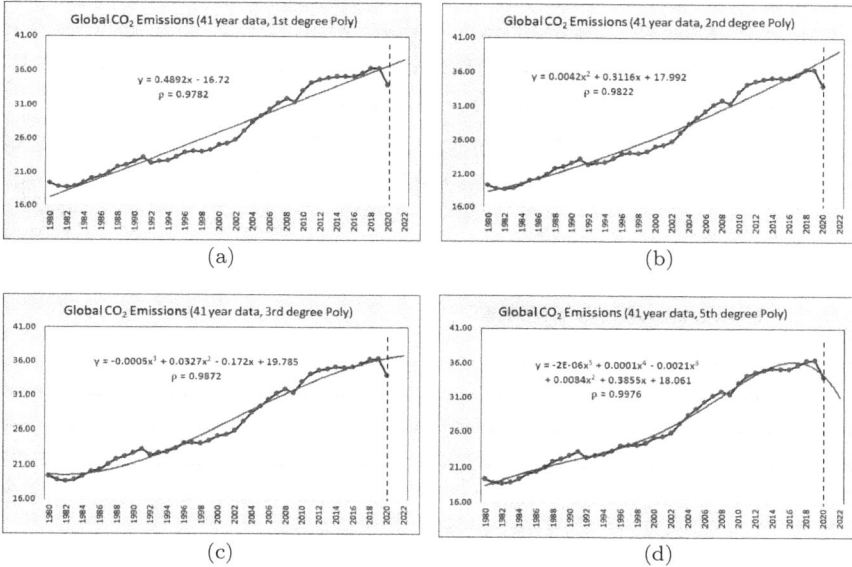

(a)   (b)

(c)   (d)

Figure 2.10: Global CO_2 emissions (function with markers), different degree polynomials (function without markers) and forecasts to the right of the vertical dashed line, derived by extrapolation of the polynomial regression function. Data from 1980 to 2020.

Let's sum up our findings:

(1) Different degree polynomials can lead to very different data representations and very different forecasts.
(2) Low degree polynomials are often biased, i.e., have a weak representation of the data with a low correlation coefficient ρ. Low degree polynomial can lead to underfitting as shown in Figure 2.4, but often represent the general trend well, see for example Figures 2.9(a) and 2.10(a). Low degree polynomials should be applied when earlier data is considered as significant as recent data.

(3) High degree polynomials often display a good data representation with a high correlation coefficient ρ. However, high degree polynomials can lead to overfitting, which can result in a questionable forecast since recent data is weighted strongly. High degree polynomials should be applied if recent data is considered more significant than earlier data.

(4) As mentioned, high degree polynomials focus on more recent data. If the recent data contains an outlier, the forecast may be distorted as seen in Figures 2.9(b)–2.9(d). An easy remedy is using a large data pool, which decreases the relative weight of the outlier as seen in Figures 2.10(a)–2.10(c).

In conclusion, statisticians can often generate a desired result by choosing a certain degree polynomial in combination with a certain time frame. So polynomial regressions are powerful tools for cheaters! Researchers ignorant of generally increasing CO_2 levels can use a high degree polynomial together with a short time frame as in Figures 2.9(b)–2.9(d), to forecast decreasing CO_2 levels. To catch potential cheaters, we should run several regressions with different degree polynomials and use a long time frame with a large data pool, which decreases the relative weight of outliers.

2.1.3. *Correlation Does Not Imply Causation!*

In this section, we will discuss the relationship between correlation and causality and how cheaters can easily misuse it.

The theory of causality has been extensively addressed in philosophy and goes at least as far back as Aristotle.[1] Causality is a central part in all natural and social sciences because humans want to know things: What caused the Big Bang? What causes the universe's expansion, which is accelerating? What causes illnesses? What causes stocks to move up and down? What causes girls to like guys, well, we (I) gave up on that one ☺.

---

[1]See Aristotle on Causality (2006), Stanford Encyclopedia of Philosophy, Dauer, F. Watanabe (2008).

The Concept of Causality

The concept of causality deals with the interaction between a primary event, the cause, and a secondary event, the effect. David Hume (1748) in his "An Enquiry Concerning Human Understanding" states eight criteria that constitute cause and effect. The critical ones are as follows:

(1) There exist two events, first, the cause, and a second, the effect.
(2) There exists a temporal order, i.e., the cause happens before or coincides with the effect. An example of a cause occurring before the effect is Newton's second law of motion $F = m\,a$, where F is the force, m is mass and a is acceleration. First a force (the cause) is applied to an object, then the objects accelerates (the effect) proportionally with the mass m. In football, if the offensive tackle (the cause) with a mass of 300 pounds accelerates and hits you (the effect), you will immediately understand Newton's second law...
(3) Effects can be probabilistic or deterministic: Smoking causing the effect cancer is probabilistic, a sun running out of hydrogen causing the effect of the sun to explode, is deterministic (in this case not only the effect but also the cause is deterministic).

Direction of Causality

Regarding the direction of causality, for two variables X and Y, we have three cases:

(1) X causes Y, the reverse not being true. Formally $(X \rightarrow Y), \neg(Y \rightarrow X)$.
(2) Y causes X, the reverse not being true. Formally $(Y \rightarrow X), \neg(X \rightarrow Y)$.
(3) X causes Y and Y causes X, termed *bidirectional causation*. Formally $(X \rightarrow Y) \cap (Y \rightarrow X)$. An example would be a stock market crash causing investor panic. It is also true that investor panic causes a stock market to crash.

"Statistics are no substitute for judgement"

(Henry Clay, Sr)

Correlation Does Not Imply Causation

Importantly, the presence of correlation does not imply that the variables in the correlation analysis are directly causally related, indirectly causally related, or causally related at all.

Let neither X cause Y, nor Y cause X. Formally $\neg(Y \to X)$, $\neg(X \to Y)$. However, even in this case of no causal relationship between X and Y, there can be a strong Pearson correlation between X and Y. The reason for the strong Pearson correlation may be the existence of a third variable W, termed confounding, hidden, or lurking variable, which causes X and Y to be related, since W causes X, and W causes Y. Formally: $\neg(X \to Y)$, $\neg(Y \to X)$, however, $(W \to X) \cap (W \to Y)$, creating a non-zero Pearson correlation between X and Y. If a third variable causes the variables of interest to be correlated without direct causation between the variables, this is termed *Spurious Relationship*. We differentiate two types:

(1) *Spurious relationship, the third variable being time*: The third variable W, which causes an effect on the variables of interest X and Y, can simply be time. As an example, in the last years we observed an increase in organic food consumption as well as an increase in Autism. Autism and organic food consumption are naturally not causally related but their correlation will be strongly positive. This constitutes the case of *nonsense-correlation*.

(2) *Spurious relationship, the third variable not being time*: An example is the default probability of Microsoft P(M) and the default probability of BMW, P(BMW), which are not *directly* causally related since they are in different countries and different industries. However, let's assume there is a worldwide recession. The default probability of Microsoft P(M) and the default probability of BMW, P(BMW) can be *indirectly* correlated via this third variable, the worldwide recession, i.e., the worldwide recession may increase both default probabilities, resulting in a positive correlation. Is this nonsense correlation? No. This

indirect correlation concept is called conditionally independent (CID) correlation approach. It is widely applied, in particular, when we have many variables to correlate. For example, when correlating the default probabilities of the 125 bonds in a Collateralized Debt Obligation (CDO) we apply the CID approach within the OFGC, the one-factor Gaussian copula model, which goes back to Vasicek (1987). For more on CID approach, see Meissner (2019, Chapter 7).

Pearson Correlation and Causation

Importantly, the Pearson correlation model does not include any methodology to determine the existence or the nature of a causal relationship between the variables in the correlation model. Therefore we have to exogenously determine the following:

(1) the existence of a direct or indirect causal relationship between X and Y; and
(2) the direction of the causal relationship (X causing Y, Y causing X, or bidirectional).

For non-experimental sciences, like economics or finance, we have to rely on theoretical analysis and empirical observations to assess the criteria (1) and (2). Only after the causality analysis has been completed and accepted, can the correlation analysis be conducted!

Let's look at an example of nonsense correlation.

Figure 2.11 shows the relationship between US spending on science, space and technology, and suicides by hanging, strangling, and suffocation.

From Figure 2.11, we see that both variables move highly correlated in time, although naturally there is no causation between the two (unless you believe that more US spending does lead to more suicides, or more suicides lead to more US spending, which you do not!).

In Figure 2.11, we have three variables, US spending, suicides, and time. To find the correlation coefficient ρ and the regression coefficient β between US spending and suicides, we plot US spending on the Y and suicides on the X axis, each point representing the

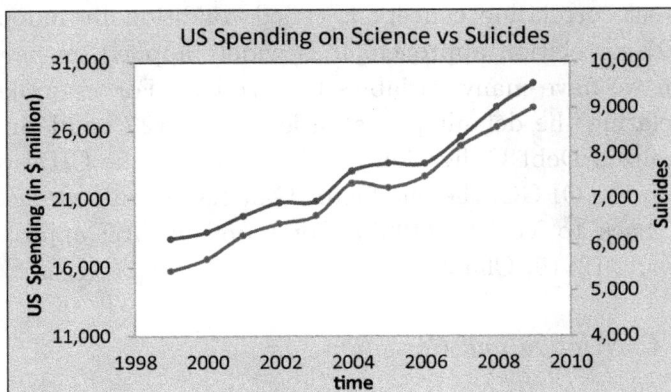

Figure 2.11: Relationship between US spending on science and technology (upper graph), and suicides by hanging, strangling, and suffocation (lower graph) in time.

Source: US Office of Management and Budget, and CDC.

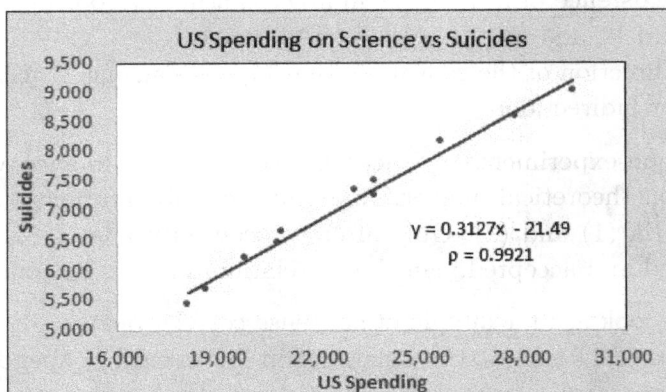

Figure 2.12: Association between suicide by hanging, strangling and suffocation, and US spending on science and technology.

Source: US Office of Management and Budget, and CDC.

US spending — suicide coordinate at a certain point in time. This results in Figure 2.12.

From Figure 2.12, we observe the very high correlation coefficient $\rho = 0.9921$. The regression coefficient $\beta = 0.3127$, as seen in Figure 2.12. This means that for every increase in US spending of \$1, on average 0.3127 more suicides occur (do keep in mind that

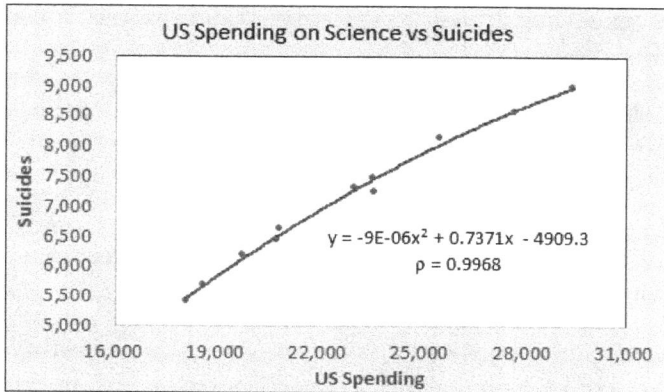

Figure 2.13: Association between suicide by hanging, strangling and suffocation, and US spending on science and technology fitted with a second degree polynomial.

this is nonsense correlation!). If we would have plotted US spending on the Y-axis and suicides on the X-axis, the regression coefficient β would be 3.1474, meaning for every increase in suicides by 1, on average 3.1474 more dollars on science is spent (this is getting more and more ridiculous!).[2]

We receive an even higher statistical correlation if we fit the data with a second degree polynomial as done in Figure 2.13.

Comparing Figures 2.12 and 2.13 we see that the correlation coefficient ρ has increased from 0.9921 to 0.9968. Figures 2.11–2.13 can be found at www.dersoft.com/nonsensecorrelation.xlsx.

For more examples of nonsense correlation, see the nice website www.tylervigen.com. If this is not enough nonsense correlation for you, Tyler Vigen even wrote a whole book on the topic, called "Spurious Correlations".

Ok, let's conclude:

The Pearson correlation model does not include an assessment on the existence or nature of a causal relationship between the variables in the analysis. Therefore, a causal analysis has to be

---

[2]The attentive reader notices that the slopes are not reciprocally, i.e., $1/3127 \neq 3.1474$. I will explain why in Appendix A.2.

performed exogenously before the correlation analysis is conducted! In particular,

(1) the existence of a direct or indirect causal relationship between X and Y; and
(2) the direction of the causal relationship (X causing Y, Y causing X, or bidirectional) have to be determined. Only after the causality analysis has been completed and accepted, can the correlation analysis be conducted!.

Naturally cheaters can exploit the correlation-causality fallacy, deriving a correlation between variables that are not causality related. For example, a researcher may falsely claim that vaccinations cause Autism, since both have increased in time. How can we catch a "correlation is not causality" cheater? Quite easy:

(a) Check if the author provides theoretical or empirical evidence for a causal relationship between the variables.
(b) Use your common sense to determine if the variables have a causal relationship.

Only if conditions (a) and (b) are satisfied, can the correlation study be considered rigorous. If not, you may have come across a statistical cheater!

2.1.4. *Kitchen Sink Regression*

To understand Kitchen Sink Regression (KSR), we have to use some math, which is so much fun ☺.

So far we looked at explaining a variable Y (for example, the AAPL stock price) with *one* variable X (for example, APPL sales). Within the regression framework, we write this as

$$\hat{Y}(X) = \hat{\beta}_0 + \hat{\beta}_1 X + \varepsilon. \tag{2.8}$$

All variables with a hat symbol ˆ are variables that are estimated in the regression analysis.

There are tons of names for $\hat{Y}$, such as dependent variable, explained variable, forecasted variable, response variable, or

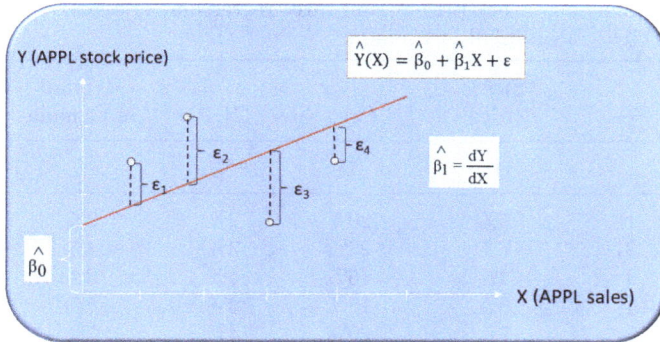

Figure 2.14: Four sample data points and the resulting regression function $\hat{Y}(X) = \hat{\beta}_0 + \hat{\beta}_1 X + \varepsilon$.

regressand, take your pick. And X also has many names such as independent variable, explanatory variable, or regressor. $\hat{\beta}_0$ is the intercept of the regression function and $\hat{\beta}_1$ is the slope of regression function, which we discussed a bit already in Chapter 1, Section 1.1.3, and with Equation (2.5). The epsilons ε are residuals, also called errors. They are the vertical difference between the data points and the regression function.[3] Graphically the regression function (2.18), resulting from four data points is displayed in Figure 2.14. Mathematically, the regression function $\hat{Y}(X) = \hat{\beta}_0 + \hat{\beta}_1 X + \varepsilon$ is derived by minimizing the sum of the squared error terms: $\min \sum_{i=1}^{n} \varepsilon_i^2$. The proof, which I am sure everyone wants to see, is at www.dersoft.com/Regressionproof.pptx.

OK, so far so good. Now let's assume we want to explain the APPL stock price Y with more than one explanatory variable X. Let's include X_2 GDP, gross domestic product. The idea is that the better the economy, the higher the APPL stock price. Let's also include a nonsense variable X_3, rainfall in Canada (which will

---

[3]Some researchers differentiate between residuals and errors. Errors are the difference between the sample data point and the unknown true population values, and residuals are the difference between the sample data point and the predicted point on the regression function, as displayed in Figure 2.14.

Table 2.2: Stock price Y and three regressors X_1 Sales, X_2 GDP, X_3 rainfall in Canada.

Year	Stock price	Sales	GDP	Rainfall in Canada
	Y	X_1	X_2	X_3
1	5%	20%	2%	3%
2	7%	23%	3%	4%
3	4%	15%	1%	3%
4	6%	18%	3%	5%
5	7%	20%	4%	7%
6	4%	16%	2%	4%
7	8%	30%	3%	4%
8	6%	25%	2%	5%
9	5%	20%	1%	6%
10	4%	18%	2%	7%
$\rho(\mathbf{Y,X}) =$		**0.8155**	**0.7536**	**0.0632**

lead us to the Kitchen Sink Regression). Naturally there is no causal relationship between rainfall in Canada and the AAPL stock price. Equation (2.8) then changes to

$$\hat{Y}(X) = \hat{\beta}_0 + \hat{\beta}_1 X_1 + \hat{\beta}_2 X_2 + \hat{\beta}_3 X_3 + \varepsilon. \qquad (2.9)$$

Let's look at a numerical example of Equation (2.9).

From Table 2.2, we observe the high correlation between the stock price Y and sales X_1. The correlation coefficient is $\rho(Y, X_1) = 0.8155$. We also find a high correlation coefficient between the stock price Y and GDP, i.e., $\rho(Y, X_2) = 0.7536$. And we find that the nonsense correlation coefficient between the stock price Y and rainfall in Canada is low at $\rho(Y, X_3) = 0.0632$. However, importantly, the nonsense variable X_3 does add some explanatory power to the regression analysis. In fact, the overall correlation coefficient, termed Multiple p, for the model with just the regressors X_1 Sales and X_2 GDP is $\rho^2(Y, X_1, X_2) = 0.8957$. Adding the nonsense variable X_3, the explanatory power of the model increases, i.e., $\rho^2(Y, X_1, X_2, X_3) = 0.9003$.

This finally brings us to the Kitchen Sink Regression problem:

> *In a Kitchen Sink Regression more and more explanatory variables are added, which, although some may be nonsense variables, typically improve the outcome of the model*

So even adding nonsense variables, with a possibly low correlation to the independent variable Y, improves the model, i.e., increases its explanatory power, for the math details see Appendix A.3. Therefore, the saying goes: "Throw everything into the regression, except the kitchen sink", since this will improve the model. The user can find Table 2.2 and the correlation coefficients at www.dersoft.com/kitch ensink.xlsx.

So if a reader finds a model with many explanatory variables, caution is advised. The more explanatory variables X, the better the model (unless the variable X has 0 correlation with Y, i.e., the correlation coefficient ρ *and* the regression coefficient $\hat{\beta}$ are 0). However, some of the new "explanatory" variables may be nonsense, nevertheless they increase the explanatory power of the model.

How to Catch Kitchen Sink Cheaters?

There is a nice concept, which addresses the kitchen sink problem. It is called the adjusted ρ^2 concept, often termed adjusted R^2, but in finance we need R for interest rate, so we use ρ^2 in this book.

The adjusted ρ^2 concept adds a penalty to the regression analysis for every newly added explanatory variable X. The equation for the adjusted ρ^2, ρ_{adj}^2, is

$$\rho_{\text{adj}}^2 = 1 - \left(\frac{(1 - \rho^2)(n - 1)}{n - k - 1} \right), \tag{2.10}$$

where k is the number explanatory variables X, n is the number of data points, and ρ^2 is the coefficient of determination, i.e., the square of the Pearson correlation coefficient ρ.

From Equation (2.10), the reader can hopefully see the positive relationship between ρ^2 and ρ_{adj}^2, (there are two minus signs in front of ρ^2), which we can write $\frac{\partial \rho_{\text{adj}}^2}{\partial \rho^2} > 0$. ∂, pronounced del, is a partial derivatives coefficient. So $\frac{\partial \rho_{\text{adj}}^2}{\partial \rho^2}$ reads: How much does ρ_{adj}^2 change, if ρ^2 changes by an infinitesimally small amount, assuming all other variable impacting ρ_{adj}^2 are constant.

From Equation (2.10), we can also observe the negative relationship between k and ρ_{adj}^2, (there are two minus signs in front of k and k is in the denominator), so $\frac{\partial \rho_{\text{adj}}^2}{\partial k^2} < 0$.

As mentioned, if we add an explanatory variable X, the explanatory power of the regression analysis measured by ρ^2 will always increase (or stay the same if ρ^2 and $\hat{\beta}$ are both zero, but this is rare), which we can write $\frac{\partial \rho^2}{\partial k} > 0$.[4] We can write the penalty of the adjusted ρ concept $\frac{\partial \rho_{\text{adj}}^2}{\partial k} < 0$. Altogether ρ_{adj}^2 will only increase if the added value of the additional new explanatory variable X is higher than the penalty, so $\left| \frac{\partial \rho^2}{\partial k} \right| > \left| \frac{\partial \rho_{\text{adj}}^2}{\partial k} \right|$. In our example of Table 2.2, the nonsense variable X_3 has low explanatory power for Y. Therefore, adding it reduces ρ_{adj}^2 from 0.8659 to 0.8505, revealing the nonsensical inclusion to the model. The reader can find these values at www.dersoft.com/kitchensink.xlsx.

So what to make of the adjusted ρ^2 concept? Can it reduce or even eliminate the kitchen sink problem? Well, it does have some nice properties:

(1) ρ_{adj}^2 only increases if the new explanatory variable is "useful". In statistical terms, ρ_{adj}^2 only increases if the new explanatory variable is significant, i.e., its addition does not increase ρ_{adj}^2 by chance.
(2) ρ_{adj}^2 will always be smaller or equal to ρ^2.
(3) It can be shown that ρ_{adj}^2 is an unbiased indicator, whereas ρ^2 is biased, therefore ρ_{adj}^2 is statistically more rigorous.
(4) If we order the explanatory variables in order of significance, ρ_{adj}^2 will increase as long as an additional explanatory variable X is useful. ρ_{adj}^2 will reach a maximum when the optimal number of

---

[4]For a mathematical explanation of this property, see Appendix A.3.

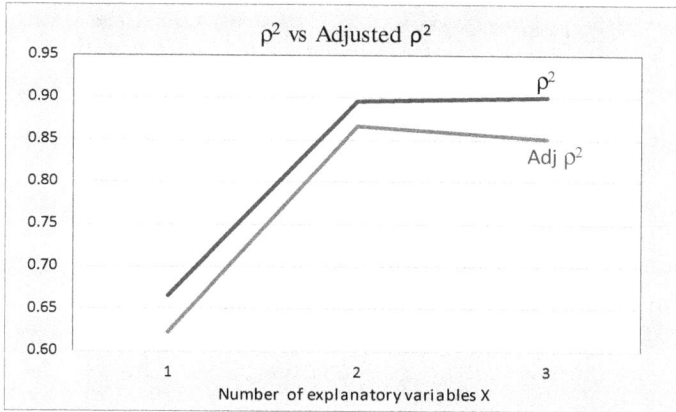

Figure 2.15: ρ^2 and adjusted ρ^2 for the data of Table 2.2.

explanatory variables X is reached, and then decrease. Pretty cool! Figure 2.15 displays the property.

From Figure 2.15, we realize that including a third nonsense variable X_3 slightly improves the explanatory power when measured by ρ^2. However, the adjusted rho square statistic ρ^2_{adj} is able to identity the third explanatory variable X_3 as nonsense, as displayed in, the decreasing ρ^2_{adj}. So the optimal number of explanatory variables X is 2, since ρ^2_{adj} reaches its maximum at 2.

What are the limitations of the ρ^2_{adj} concept?

(1) ρ^2_{adj} can, contrary to ρ^2, get negative. As we learned in 4th grade, a squared number cannot be negative. So the result is nonsensical (unless you go deeper into math, where we call the negative of squared numbers "imaginary numbers" which do have real-world applications). For example, the parameter combination of $\rho^2 = 0.1, n = 100$ and $k = 15$, results in $\rho^2_{\text{adj}} = -0.0607$ as the reader can verify at www.dersoft.com/kitchensink.xlsx.

(2) Like many statistical concepts, ρ^2_{adj} suffers from the "increasing data fallacy", meaning the more data n you use, the better the result, in this case, the higher ρ^2_{adj}. From Equation (2.10) $\rho^2_{\text{adj}} = 1 - \left(\frac{(1-\rho^2)(n-1)}{n-k-1} \right)$, it is actually not so easy to see what impact

n has on ρ^2_{adj}, since n is in the numerator and denominator. So we have to use, sorry, good old calculus, which I did in cell F11 in www.dersoft.com/kitchensink.xlsx. For the sensitivities of ρ^2_{adj} with respect to n, k and ρ^2, see www.dersoft.com/adjustedrhod erivatives.pdf.

In conclusion, kitchen sink cheaters, who claim that their regression model with tons of explanatory variables X can explain the dependent variable Y well, can be caught with the adjusted ρ^2 concept. Just calculate it. This will reveal (1) the true explanatory power of the model, (2) the nonsense variables, and (3) the optimal number of regressors X.

2.1.5. *Multicollinearity Can Distort Individual Regressors*

The world is complex. So when we run a regression to explain what causes a phenomena like Global Warming, Cancer, a Recession, or a Stock price change, we typically use several variables to explain the phenomena. We already did this in Table 2.2. The problem now is that these explanatory variables, also called regressors, can be correlated amongst themselves. So we can define

> *Multicollinearity exists when two or more of the explanatory variables X are correlated*

Let's look at an example where Y are the new infections of a Pandemic. X_1 is number of people who socially distance, and X_2 is number of people who are vaccinated. And we have a third variable X_3 which is ice cream consumption in Milwaukee. Our objective is to explain the new infections Y in time. We formulate the regression problem with the same equation, which we used in the previous Kitchen Sink Regression:

$$\hat{Y}(X) = \hat{\beta}_0 + \hat{\beta}_1 X_1 + \hat{\beta}_2 X_2 + \hat{\beta}_3 X_3 + \varepsilon. \qquad (2.9)$$

Multicollinearity exists if the regressors X_1, X_2 and X_3 are correlated. Why is that a problem? Well, there are two:

(1) While the overall regression coefficients ρ and $\hat{\beta}_1$ of the regression analysis are not distorted, the explanatory power of the individual regressors is distorted since they are correlated. So a statistical cheater may be able to show that ice cream consumption in Milwaukee has a high impact on new infections Y. Or, more seriously, that a noneffective drug has a positive impact on an illness. We call this a false positive or a type I error. Vice versa, distorted regressors may indicate that a regressor X has no impact on the dependent variable Y, when it has. We call this a type II error, a false negative.

(2) In fact, highly correlated regressors X typically result in high standard errors. This can lead to falsely accepting the null hypothesis, the type II error. This is actually a "bad error", at least in medicine, since you are telling the patient he is healthy when he is not. A type I error in medicine is diagnosing a patient with an illness, but he is actually healthy. However, this is not as bad a type II error, because additional tests will reveal that the patient is healthy.

How to Detect Multicollinearity?

Let's first see how to determine if multicollinearity is present. That is actually pretty easy. Let's look at an example of the data in Table 2.3. We have the infection rate of a virus, say, maybe COVID-19, and potentially influencing factors X_1 and X_2 and a nonsense variable X_3.

From Table 2.3, we observe the very high negative correlation between Y and the regressors $\rho(Y, X_1) = -0.9639$, $\rho(Y, X_2) = -0.9876$. The correlation coefficient between Y and the nonsense variable X_3 is also high at $\rho(Y, X_3) = -0.9000$.

So how can we determine if multicollinearity is present? First, we can look at the data. From Table 2.3, we are already quite suspicious that the regressors X are correlated, since they move largely up and down together. To verify this suspicion, we run a correlation analysis between the regressors. We get Table 2.4.

Indeed, Table 2.4 tells us that the regressors X are highly correlated! Is this exciting or what?

Table 2.3: Infection rate Y, its three regressors X_1, X_2, and X_3 and the correlation coefficients.

Week	Infection Rate	Individuals who socially distance	Individuals who are vaccinated	Ice cream consumption in Milwakee
	Y	X_1	X_2	X_3
1	12%	10%	5%	2%
2	11%	11%	6%	4%
3	10%	8%	7%	8%
4	9%	12%	8%	9%
5	8%	13%	9%	10%
6	6%	16%	11%	9%
7	5%	18%	12%	10%
8	4%	20%	14%	11%
9	3%	22%	16%	12%
10	2%	24%	18%	13%
$\rho(Y,X) =$		**−0.9639**	**−0.9876**	**−0.9000**

Table 2.4: Correlation coefficient ρ matrix of the regressors X.

	X_1	X_2	X_3
X_1	1	0.9734	0.7798
X_2	0.9734	1	0.8760
X_3	0.7798	0.8760	1

Consequences of Multicollinearity

(1) Distorted Regressors

Importantly, multicollinearity does not violate any assumptions of the Pearson regression model. As a consequence, the analytical results of a regression analysis as a whole are unbiased.

However, since the individual regressors are highly correlated, it is difficult to determine their individual impact. Therefore the analytical power of the individual regressors can be distorted. Let's show that with our example. Table 2.5 shows the results of the multivariate regression $Y = f(X_1, X_2, X_3)$.

Table 2.5: Regression output for the data in Table 2.3.

SUMMARY OUTPUT

Regression Statistics

Multiple R	0.9933
R Square	0.9867
Adjusted R Square	0.9800
Standard Error	0.0049
Observations	10

ANOVA

	df	SS	MS	F	Significance F
Regression	3	0.0109	0.00362	148.1583	5.14E-06
Residual	6	0.0001	2.4E-05		
Total	9	0.0110			

	Coefficients	Stand error	tStat	P-value
Intercept	**0.1657**	**0.0074**	**22.3395**	**0.0000**
X Variable 1	**−0.2919**	**0.1756**	**−1.6626**	**0.1475**
X Variable 2	**−0.2312**	**0.2833**	**−0.816**	**0.4457**
X Variable 3	**−0.2982**	**0.1326**	**−2.2493**	**0.0655**

The upper non-bold data are the overall regression statistics, which are not distorted by multicollinearity. However, the lower parameters in bold are the results for the individual regressors, which can be distorted by multicollinearity. Are they in our example?

Well, let's look at the p-value (not to be confused with the correlation coefficient ρ), displayed in the last four rows of Table 2.5. Loosely put, the lower the p-value the higher the significance, i.e., the higher the impact a regressor X has on Y.[5] The p-value is lowest for the nonsense variable X_3. That can't be right, and it is not, since we already calculated the individual correlation coefficients for each regressor in Table 2.3 and found that the nonsense variable X_3 has the lowest correlation with Y. The t-statistic (which can be inferred from the p-value so we don't really need to calculate it) is

---

[5]The exact interpretation of the p-value depends on the mathematical framework (Bayesian or Frequentist) and is test-dependent. It is an ongoing discussion in the statistics community. For one of many sites discussing the p-value see https://www.researchgate.net/post/P-value-and-Type-1-Error-confusing-concept.

also misleading, since it is the highest (in absolute terms), so shows the highest significance of X_3.

(2) Increased standard errors which may lead to type II errors

So far we discussed that multicollinearity can distort the impact of individual regressors. Let's now show that multicollinearity often leads to an increase in the standard error, which in turn can lead to a type II error. To do this we need some math, sorry. Let's look at Equation (2.11).

$$\text{se}_{\beta j} = \sqrt{\frac{\sum_{i=1}^{n}(y_i - \hat{y}_i)^2}{\sum_{i=1}^{n}(x_i - \hat{x}_i)^2(1 - \rho_j^2)(n - k)}}, \qquad (2.11)$$

where $\text{se}_{\beta j}$ is the standard error of the regression coefficient β of the regressor j, ρ_j^2 is the coefficient of determination of the jth regressor, which is regressed (input as a dependent variable) against the other regressors, n is the number of data points, and k is the number of regressors X.

The standard error of the regression coefficient β of the regressor j, $\text{se}_{\beta j}$, of Equation (2.11) looks a bit nasty. Here is some intuition: Loosely put, the lower the standard error, the more reliable is the regressor j in explaining Y. We can see this from the numerator, which gives the (squared) deviation of the data points Y_i with respect to the points on the regression function $\hat{y}_i$. The lower this deviation, the more reliable our test, so the lower the standard error.

And we can also look at the denominator. A higher deviation of the regressor data points x_i with respect to the data points on the regression function $\hat{x}_i$ decreases the standard error $\text{se}_{\beta j}$. Why? Well, intuitively, the higher this deviation, the more "stretched" are the data points on the x-axis. Therefore, the lower is the slope of the regression function, i.e., it is quite stable and less likely to change from positive to negative or vice versa.

And we also observe from Equation (2.11) that the more data points n we have, the more reliable our test, so the lower the standard error.

Ok, let's show why a distorted standard error can be a problem. As we see from Equation (2.11), if the correlation between j and the other regressors is high, i.e., we have a high ρ_j^2, the standard error $se_{\beta j}$ will be high (there is minus sign in front of ρ_j^2, and ρ_j^2 is in the denominator). These distorted high standard errors may lead to the failure of rejecting the false null hypothesis, so result in a too high standard error $se_{\beta j}$, a type II error. Put simply, due to multicollinearity, the model fails to recognize the true, strong explanatory power of the regressor. As discussed above, in medicine this a "bad error" because you are telling patients they are healthy, when they are not. We also observe from Equation (2.11) that for total multicollinearity, i.e., $\rho_j^2 = 1$, the regression falls apart, i.e., the standard errors cannot be evaluated.[6]

What to Do If Multicollinearity is Present?

As discussed, a statistical cheater can run a regression with highly correlated regressors and falsely claims that some are significant when they are not, as in the example of Table 2.3. He or she can also observe multicollinearity-inflated standard errors, which can lead to a type II error, thus claiming a regressor has no significant impact on Y, when it has. How can we catch him or her? Pretty easy:

(1) We should first find out if multicollinearity is actually present. We can do this by correlating the regressors with each other to find the degree of multicollinearity, so calculate $\rho(X_1, X_2)$, $\rho(X_1, X_3)$, $\rho(X_2, X_3)$, as we did in Table 2.4. We could also run a regressor with respect to more than one regressor, so for our example calculate $X_1 = f(X_2, X_3)$, $X_2 = f(X_1, X_3)$, $X_3 = f(X_1, X_2)$ to see the combined impact of the regressors on another regressor.

(2) To find out the individual impact of a single regressor we should run the regression with just that single regressor. We did that in Table 2.3 and reported the individual correlation coefficients

---

[6]For a nice paper discussing further properties of multicollinearity see Stephen Voss, "Multicollinearity" *Encyclopedia of Social Measurement*, 2004.

$\rho(Y,X_1)$, $\rho(Y,X_2)$, $\rho(Y,X_3)$ for every regressor. This individual impact may be compared with the impact of the regressor in the multivariate regression. A difference may indicate a distorted regressor impact due to multicollinearity.

(3) We could increase the data pool or run the data out-of-sample, which may decrease multicollinearity and give better estimates of the individual regressors.

2.1.6. *Heteroskedasticity to Manipulate Standard Errors*

While difficult to pronounce and spell, heteroskedasticity (sometimes heteroscedasticity), is a simple concept. It comes from the Greek "hetero" meaning different and "skedasis" meaning dispersion. A picture says 1,000 words, so let's look at heteroskedasticity graphically first:

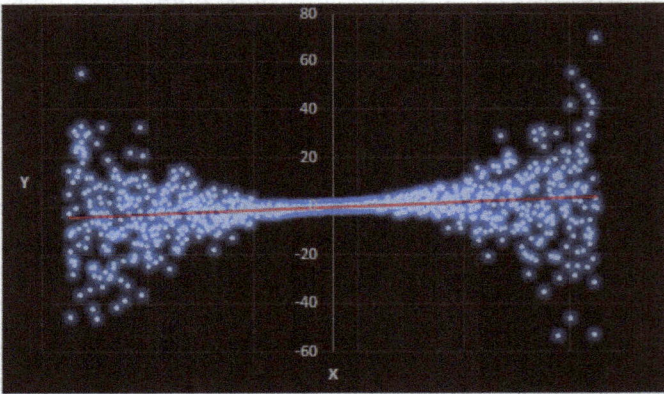

Figure 2.16: Data points and resulting regression function.

We are interested in the error terms, which are the vertical differences of the data points to the regression function. In the example of Figure 2.16, the dispersion of the error terms is a function of X: The higher and lower X, the higher the vertical dispersion of the data points, so the higher the error terms. So we can define

---

Heteroskedasticity exists if the error terms are not constant, but vary with X

Ok, now we have to do a bit math, which I am sure the reader will so enjoy ☺. We will measure the dispersion of the error terms with the variance Var, which is the square of the standard deviation, you may look at Equation (2.2). Then heteroskedasticity exists if

$$\text{Var}(\varepsilon_i|X) = f(X). \tag{2.12}$$

Equation (2.12) reads: The variance of the error terms ε_i conditionally on X, $\text{Var}(\varepsilon_i|X)$, is a function of X, $f(X)$, therefore ε_i varies with X.

Conversely, homoskedasticity exists if the variance of the error terms is constant in X:

$$\text{Var}(\varepsilon_i|X) = \sigma^2, \tag{2.13}$$

where σ^2 is a constant, representing the constant variance of the error terms.

Homoskedastiticy is one of the three Gauss-Markov assumptions, the other two being that the error terms have zero expectation: $E(\varepsilon_i|X) = 0$ and are uncorrelated: $\text{Cov}(\varepsilon_i, \varepsilon_j) = 0 \ \forall \ i, j, i \neq j$. Gauss and Markov show that the three assumptions result in the best linear unbiased estimator (BLUE) for the coefficients $\hat{\beta}_0$ and $\hat{\beta}_1$ of the OLS (ordinary least squares) approach.

Consequences of Heteroskedasticity

Why is heteroskedasticity a problem? First, let's see what is not a problem. When deriving the optimal values of the regression parameters $\hat{\beta}_0$ and $\hat{\beta}_1$ of the regression function $\hat{Y}(X) = \hat{\beta}_0 + \hat{\beta}_1 X + \varepsilon$ we differentiate the squared error terms ε_i^2 with respect to $\hat{\beta}_0$ and $\hat{\beta}_1$. In this process, it is irrelevant if the variance of error terms vary in X, i.e., if the error terms are heteroscedastic (proof is at www.dersoft .com/heteroskedasticityproof.pdf). Hence the regression parameters $\hat{\beta}_0$ and $\hat{\beta}_1$ are robust with respect to heteroskedasticity.

However, the conventional standard error term of the regression coefficient $\hat{\beta}, \text{se}(\hat{\beta})$,

$$\text{se}(\hat{\beta}) = \sqrt{\frac{\sigma_\varepsilon^2}{\sum_{i=1}^{n} (x_i - \overline{x})^2}}, \tag{2.14}$$

is a function of the variance of the error term σ_ε^2, as we can see from Equation (2.14). Therefore the standard error $\text{se}(\hat{\beta})$ is not heteroskedasticity robust, for a proof, see www.dersoft.com/heteroskedasticityproof1.pdf. The standard error is an input for inference tests as the t-test, z-test or F-test. For example, the t-statistic for the regression parameter $\hat{\beta}$ is $t_{\hat{\beta}} = \frac{\hat{\beta}-\beta}{\text{se}(\hat{\beta})}$. Hence inference tests which include the standard error are non-heteroskedasticity robust and can give false results such as falsely rejecting the null hypothesis (type I error) or falsely accepting the null hypothesis (type II error).

How to Deal With Heteroskedasticity?

Similar to the limitation of multicollinearity, the first thing to do when dealing with heteroskedasticity is to find out if it is present. We can investigate the scatterplot visually to get a good idea whether the data is heteroskedastic. In Figure 2.16, we could already visually observe that heteroskedasticity is present. Numerous statistical tests for heteroskedasticity exist, for example, the Breusch–Pagan test, which tests the null-hypothesis that error variances are equal. The White test, a special case of Breusch–Pagan, which can handle non-normal errors, and other tests such as the Park test or the Glesjer test can be applied.

If heteroskedasticity does exist, we can apply heteroskedasticity robust standard errors, which are typically credited to Halbert White (1980). However, some groundwork was laid by Eicker (1967) and Huber (1967). The White heteroskedasticity robust standard error statistic $\text{se}(\hat{\beta})_W$, also called Eicker–Huber–White standard error, does not apply the variance of the error term σ_ε^2, but the error terms ε_i themselves in the test statistic. Hence Equation (2.14) changes to

$$\text{se}(\hat{\beta})_W = \sqrt{\frac{\sum_{i=1}^{n} \varepsilon_i^2 (x_i - \overline{x})^2}{\left(\sum_{i=1}^{n} (x_i - \overline{x})^2\right)^2}}. \tag{2.15}$$

One question remains: If heteroskedasticity-robust standard errors exist (Equation (2.15)), why bother with non-heteroskedasticity-robust standard errors of Equation (2.14)? The answer is that if homoscedasticity holds and the errors are normal, then the t-statistic applying Equation (2.14) will be *exactly* student-t distributed. However, heteroskedasticity-robust standard errors may not be exactly student-t distributed, especially for small sample sizes. This may lead to biased inference statistics. Therefore some studies derive and report both heteroskedasticity-robust and non-heteroskedasticity-robust standard errors.

Conclusion Heteroskedasticity

What to make of all this? Well, in the example of Figure 2.16, we can see that the variance of the error term σ_ε^2 can be stated lower, which it is close to $X = 0$, or stated higher, which it is for low and high values of X. So when putting the variance of the error term into Equation (2.14), a lower or higher standard error se$(\hat{\beta})$ can result. A standard error which is biased lower can lead to a type I error, a false positive, i.e., falsely claiming that a regressor X has an significant impact on Y, when it does not. A standard error which is biased higher, can lead to a type II error, a false negative, i.e., claiming that a regressor X has no significant impact on Y, when it does.

A remedy for the distorted standard errors due to heteroskedasticity is to apply heteroskedasticity robust standard errors. However, they have the drawback of being imprecise, since they are not exactly student-t distributed, especially for small sample sizes. What to do, what to do? Well, to catch statistical cheaters who derive distorted standard error terms due to heteroskedasticity, we can calculate both conventional standard errors with Equation (2.14) and heteroskedasticity robust standard errors with Equation (2.15). This will give us valuable information on the standard error.

Summary and Conclusion: How to Catch Regression Cheaters — Pretty Easy

In this chapter, we discussed flaws in Regression analysis, which a statistical cheater can exploit. Regression analysis has two main objectives:

(a) To find dependencies between variables, for example, which drug can cure an illness, or what makes a stock price rise.
(b) In finance, we often deal with times series. In this case, we can extrapolate the regression function and try to forecast a variable such as a stock price.

We discussed six flaws within Regression analysis:

(1) The most widely applied Regression model, the Pearson Regression model, derives only the *linear* relationship between variables. Therefore, if a statistical cheater evaluates nonlinear relationships with the Pearson model, the true nonlinear relationship between variables is typically underestimated or not found at all, see, for example, Figures 2.4–2.6. So the reader should look at the data points. If they are nonlinear, the Pearson model is typically not a good model to derive dependencies between the variables.
(2) The linear limitation of the Pearson regression model can be addressed by deriving a nonlinear regression function. We call this polynomial or nonlinear fitting. However, this can still lead to two problems:

- Underfitting, which is fitting a too low degree polynomial through the data, which does not represent the nonlinear relationship between the variables well.
- Overfitting, which is fitting a too high degree polynomial through the data. In this case, the long-term relationship between the variables is not represented well.

Therefore, statistical cheaters can fit a certain degree polynomial to derive a desired outcome, for example, claim that CO_2 emission are declining (see Figures 2.9(b)–2.9(d)). Generally, if older data are considered equally relevant as recent data, a low degree polynomial, although it results in a lower correlation coefficient than a higher degree polynomial, is sensible. If more recent data is considered more relevant, a higher degree polynomial is sensible. If an outlier is distorting the results, a simple remedy is to use a large data pool, which reduces the relative impact of the outlier.

(3) The Pearson model does not include any methodology to determine whether the variables are actually causally related. So variables can be correlated without being causally related. We call this Spurious Correlation or Correlation $\neq$ Causation. So a cheater can find a strong correlation between variables that are not causally related such as Autism and organic food consumption, or CO_2 emissions and longevity. A remedy is (a) see if the author has verified the causality between the variables, either logically or empirically, and (b) use your own knowledge and common sense to determine if the variables are causally related.

(4) A statistical cheater can exploit the "Kitchen Sink Property" of a regression model. It means that the more regressors X you add to the model, the better the outcome. This is true even if the additional variable is nonsense, like a kitchen sink, and has a low correlation to the variable that we want to explain Y. There is an easy remedy though: The adjusted ρ^2 concept, often called adjusted R^2. It adds a penalty to the regression model for every added regressor X and only improves the outcome of the model if the additional regressor X is "useful", in statistical terms if its addition to the model improves the model by more than just by chance.

(5) Multicollinearity means that the regressors X are correlated. This does not distort the overall regression result, however, it can distort the impact of individual regressors since they are correlated. So a cheater may be able show that rainfall in Milwaukee has a certain impact on COVID-19 infections, or vice versa that

a significant variable such as vaccination has little impact on the infection rate, when it has a high impact. Fortunately, the problem of multicollinearity can be easily solved: We can run the regression for each regressor X individually to determine their individual impact. We can also increase the data pool or run the data out-of-sample, which may decrease multicollinearity and give better estimates of the individual regressors.

(6) Heteroskedasticity is difficult to spell and pronounce, however, it is a simple concept. It just means that the error terms are not constant, but vary with X (see Figure 2.16). This can cause problems with standard errors, so can lead to a type I error, which is a falsely rejecting the null hypothesis, thus claiming the regression analysis is significant when it is not. It can also lead to type II errors, therefore, falsely accepting the null hypothesis, so not finding the true explanatory power of the regression analysis. The remedy to heteroskedasticity are heteroskedasticity robust standard errors. They can typically catch cheaters who exploit heteroskedasticity distortions.

2.2. Inferential Statistics

So far we found how a statistical cheater can exploit flaws in Descriptive Statistics. As the term implies, Descriptive Statistics mainly just describes phenomena, for example, the association between a vaccination and the infection rate. This can be done by deriving the correlation coefficient ρ and the regression coefficient β, which we did in Section 2.1.

We now go a step further and draw conclusions from the descriptive analysis, called Inferential Statistics, which analyzes if an association happens just by chance or if it is "significant" as we so say in statistics. Fascinating stuff... So we can define

---

Statistical Inference draws conclusions for associations. In particular, it analyzes if an association just happens by chance, or is significant.

Hypothesis Testing

Let's look at one of the most critical fields of inferential statistics, hypothesis testing. Let me explain it with an example, if I may ☺. We have three steps:

(1) Formulating the Hypotheses

First, we have to formulate what we want to do. Let's take a simple example. You are in Las Vegas (if you go there once in life that's enough), and the dice in craps keeps rolling a lot of sixes. You are getting suspicious. So we formulate the *Null-hypothesis* H_0

$$H_0 : \text{the dice is fair and not manipulated}$$

and the research hypothesis, which is typically called *Alternative hypothesis* H_1:

$$H_1 : \text{the dice is manipulated to roll sixes}$$

(2) Choosing a Confidence and Significance Level

We have to choose a *confidence level*, which is arbitrary. Typical confidence levels are 95%, 99% or 99.9%. Let's choose 95% for our test, which means that we are 95% confident that our test results are correct. From the 95% confidence level, we get a $1 - 0.95 = 0.05$ or 5% *significance level.*

(3) Testing

Now we have to conduct tests to see whether the dice is manipulated. We steal the dice from the casino, go home and roll the dice twice. Two sixes appear. Hmmm, suspicious. But how can we be sure? Well, the probability of rolling the dice twice and getting two sixes is $1/6 \times 1/6 = 1/36$ or 2.78%. So the probability is very low. We compare this low probability with our significance level and find $2.78\% < 5\%$. Therefore, we conclude that the dice is manipulated to roll sixes.

There are a couple of issues with our test though:

(a) Are we sure that the dice is manipulated? No. We are just 95% confident that our result is correct. If we are unlucky, the dice is fair and the two sixes appeared just by chance! This is what we call at type I error: We falsely reject the Null hypothesis that the dice is fair, and falsely accept the Alternative Hypothesis that the dice is manipulated. We call this a false positive.

(b) It can also happen that we roll a 3 and a 6. Now what? Well, the probability of rolling *one* six in two trials is 10/36 or 5/18 or 27.78% (if you don't believe me just write down the 36 possibilities). Since our test result is much bigger than our significance level, i.e., 27.78% > 5%, we conclude that the dice is not manipulated to roll sixes. What is the problem here? Well, it can be that the dice is manipulated to roll sixes often, but not always. So our conclusion that the dice is fair is false. So we are falsely accepting the null-hypothesis that the dice is fair, and are falsely rejecting the alternative hypothesis that the dice is manipulated. We call this a false negative or Type II error. As we already mentioned in Section 2.1.5, in medicine this is a bad error because you are telling the patient he is healthy, when he is sick.

What to do to solve the problem of a type I and type II error? Pretty easy: We just have to do a lot of testing. If we let our unpaid intern roll the dice 1,000 times, we have a fairly large sample to find out how many sixes more are rolled with our dice compared to a fair dice, where the probability of rolling sixes is 1/6. So naturally, as with most statistical tests, the larger the sample size, the more reliable the test results. So when the reader finds a statistical test with a small sample size, caution is advised. The researcher may be a statistical cheater.

The Good Old z-Test and How to Cheat With It

In Chapter 1, Section 1.1.5, we used the z-score Equation (1.2): $z = \frac{x-\mu}{\sigma}$ to standardize data. For hypothesis testing, we expand the z-score concept and write

$$z = \frac{\overline{x}-\mu}{\sigma/\sqrt{n}}, \tag{2.16}$$

where $\overline{x}$ is the sample mean, μ is the (typically unknown) population mean, σ is the (typically unknown) population standard deviation, and n is the sample size.

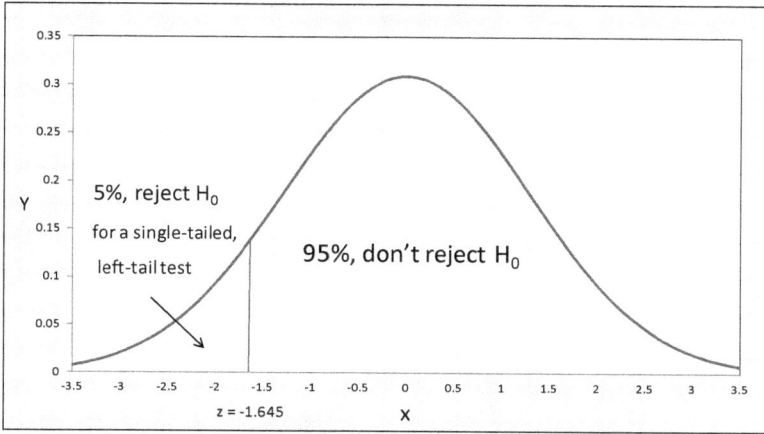

Figure 2.17: Normal distribution with a 5% significance level and the resulting critical z-score of $z = -1.645$. If the test result is lower than -1.645, the null hypothesis H_0 is rejected.

If the z-score of Equation (2.16) is higher than a critical positive value or lower than a critical negative value, we conclude that the test result is statistically significant. This can be best explained with a graph. Figure 2.17 shows a left tail-test with a 5% significance level.

We conduct a single-tailed, left-tail test if we want to find out if a certain action leads to a significantly lower result. For example, we want to see if a vaccine leads to significantly lower infection rates, or wearing a helmet leads to significantly lower motorcycle deaths. If a test results in a z-score of ≤ -1.645, we call this "statistically significant". In Excel, a left tail 5% significance z-score is derived as $= \text{normsinv}(0.05) = -1.645$. It is called norminv because we are inverting a y-axis value (0.05) of a cumulative normal distribution to derive the x-axis value (-1.645).

In Python, the syntax is

```
INPUT
from scipy.stats import norm
print(norm.ppf(0.05))

OUTPUT
=-1.644853626951
```

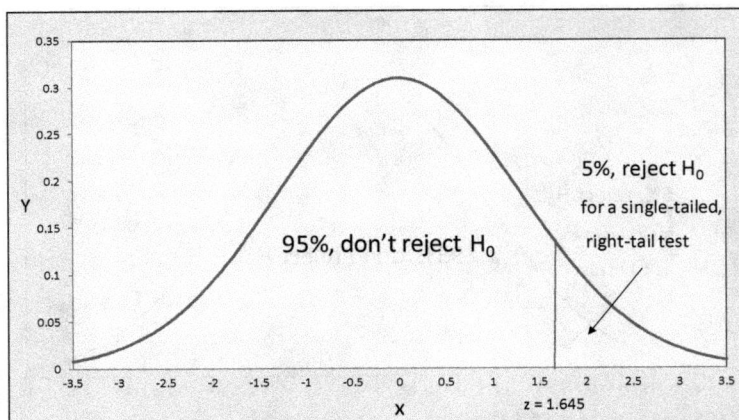

Figure 2.18: Normal distribution with a 5% significance level and the resulting critical z-score of $z = 1.645$. If the test result is higher than 1.645, the null hypothesis H_0 is rejected.

This is pretty simple stuff. But if the reader wants to run this Python code in the cloud, here is the link: https://replit.com/@GunterMeissner/z-score#main.py. For a graph explaining how the z-score is derived, see Appendix A.4.

Vice versa, we conduct a single-tailed, right-tail test if we want to find out if a certain action leads to a significantly higher result. For example, we want to see if a hedge fund's trading has significantly outperformed the market, or more if education leads to significantly higher income. This is shown in Figure 2.18.

As shown in Figure 2.18, a test result with a z-score of $z \geq 1.645$ falls into the right 5% tail of the standard distribution. We call this result "statistically significant". In Excel, a right tail 5% significant z-score is derived as $= \text{normsinv}(0.95) = 1.645$. In Python, the syntax is

```
INPUT
from scipy.stats import norm
print(norm.ppf(0.95))

OUTPUT
=1.644853626951
```

2.2.1. *Increasing the Sample Size n to Increase Significance*

How to cheat with the z-test? Now that's an easy one. Let me explain it with an example.

Example 2.1: *Hedge fund manager Mr. Wonderful claims that in the last 150 days he has outperformed the market, i.e., his portfolio return is significantly higher than that of the market, represented by the S&P 500. Let's assume that he has done quite well: His return was 12% in comparison to the S&P 500, whose return was 10%. The critical question is: Was Mr. Wonderful just lucky and his outperformance happened by chance, or is his outperformance "statistically significant"? To answer this question, we calculate the z-score.*

We have $\bar{x} = 12\%$. We know how the market performed, so we know the population mean $\mu = 10\%$. Let's assume we also know the population standard deviation, so the standard deviation of the daily market return. It is $\sigma = 15\%$. The time frame of observation is $n = 150$ days. Throwing this all into Equation (2.16) we get

$$z = \frac{\bar{x} - \mu}{\sigma/\sqrt{n}} = \frac{0.12 - 0.10}{0.15/\sqrt{150}} = 1.633.$$

If we apply a significance level of 5%, we realize that the z-score of 1.633 is lower than the critical z-score of 1.645. Therefore, we conclude that the outperformance of Mr. Wonderful is not statistically significant. The reader can find this result at $www.dersoft.com/z$ score.xlsx, cell G6.

Furious, that his outperformance is not statistically significant, Mr. Wonderful manipulates the z-score result (my apologies to the Mr. Wonderful from shark tank, who is surely an honest man ☺). The hedge fund manager Mr. Wonderful increases the sample size n from 150 to 153 days. Let's assume this does not change his performance of $\bar{x} = 12\%$, the market return of $\mu = 10\%$ and the standard deviation of the daily market return of $\sigma = 15\%$. The new z-score is now

$$z = \frac{\bar{x} - \mu}{\sigma/\sqrt{n}} = \frac{0.12 - 0.10}{0.15/\sqrt{153}} = 1.649.$$

Since this z-score is higher than the critical z-score of 1.645, it falls into the critical 5% right tail. Therefore the outperformance of Mr. Wonderful is now statistically significant! The reader may check this result at www.dersoft.com/zscore.xlsx, cell G6 (change cell C8 to 153).

Conclusion: One way to manipulate the z-score is simply increasing the sample size. We can call this the "Increasing Data Fallacy". We already experienced a similar statistical property with the "Kitchen Sink Regression" in Section 2.1.4: The more regressors we use in the regression analysis, the better the regression result.

What to do to catch a sample size cheater? It's not so easy. There is actually a paradox: The bigger the sample size, the more rigorous the statistical test. However, the bigger the sample size, the better the statistical result. The observer should watch for either extreme: A very low sample size means little statistical rigor. In fact, for a very low sample size, it may be impossible to achieve significance at all: If you want to test if a dice is manipulated to roll sixes and you only roll it once, you cannot achieve statistical significance: Even if a six is rolled this probability is $1/6 = 0.1667$, which is smaller than the critical 5% z-score of 1.645! Vice versa, the reader should be suspicious for very large samples sizes, which typically inflate test results.

> *"The life expectancy of a single person is quite uncertain. The average life expectancy of many is not."*
>
> (Elizur Wright)

2.2.2. *Manipulating the Population Mean μ*

In the previous Example 2.1, we worked with historical financial data. In this case, the population mean and population standard deviation are often known. However, for many tests, both parameters are not known and have to be estimated. This is a great cheating opportunity! Let's look at an example:

Example 2.2: *Jeff is very proud of his height and wants to prove to his girlfriend that he is significantly taller that the average US male.*

Jeff is 180 cm tall. (I am using centimeters since it is my goal in life to introduce the metric system to the US which is impossible). To use the z-score Jeff has to estimate the average US male height, which is not precisely known. Jeff asks 30 male students and calculates the average height as 177 cm. In addition, Jeff has to estimate the US population male height standard deviation. After some research he assumes it is 15. Jeff throws his data into the z-score, which comes out as

$$z = \frac{\bar{x} - \mu}{\sigma/\sqrt{n}} = \frac{180 - 177}{15/\sqrt{30}} = 1.095. \qquad (2.17)$$

Jeff applies a significance level of 5% and realizes that the z-score of 1.095 is lower than the critical z-score of 1.645 (compare Figure 2.18). Therefore his height is not significantly higher than the average US male, so Jeff is too short to show off.

Frustrated, Jeff decides to manipulate the estimated US male height μ. He includes some shorter students in his sample of 30 students and finds that now the average height is 175 cm. Putting this new estimated US population height into the z-score equation results in

$$z = \frac{\bar{x} - \mu}{\sigma/\sqrt{n}} = \frac{180 - 175}{15/\sqrt{30}} = 1.826.$$

The reader can find this result at www.dersoft.com/zscore.xlsx, cell G15.

Since this z-score is now higher than the critical z-score of 1.645, it falls in the critical 5% right tail (see Figure 2.18). Jeff shows his statistically significant height difference compared to the average US male to his girlfriend who has no clue what he is talking about but is very impressed by his academic knowledge (as Penny in Big Bang Theory).

In conclusion, the population mean is often not known and has to be estimated. A statistical cheater can manipulate the estimation to derive a desired result. What to do? The reader should check if the researcher has performed a rigorous estimation of the population mean, either empirically or theoretically. The reader can also research herself if the population mean is sensible.

2.2.3. *Manipulating the Population Standard Deviation Parameter, the "Nuisance Parameter"* σ

In the previous section, Jeff manipulated the estimation of the population mean μ to derive the desired significance. However, the population mean can often be fairly well estimated, since it is just a mean. In the previous example, there is data on the mean height of US males.

Let's assume Jeff's girlfriend found that the average height of US males is actually 177 cm, not 175 cm. Jeff is busted, proven too short. However, he has one more trick up his sleeve. It is the standard deviation σ of the US male height. This is more difficult to check since data is hard to find. So even if Jeff uses the correct estimate of the US male average height of 177 cm, he can manipulate the standard deviation σ. Jeff decides to decrease the US male height standard deviation estimation σ from 15 to 8.5. Hence Equation (2.17) changes to

$$z = \frac{\bar{x} - \mu}{\sigma/\sqrt{n}} = \frac{180 - 177}{8.5/\sqrt{30}} = 1.933. \tag{2.18}$$

The reader can find this result at www.dersoft.com/zscore.xlsx, cell G24. Since the z-score of 1.933 is higher than the critical 5% z-score of 1.645 (compare Figure 2.18), Jeff is now statistically significantly taller than the average US male. Yay!

Since the estimated population standard deviation σ is easy to manipulate and hard to verify, it is also called a "nuisance parameter". What to do? Just as with the estimated population mean μ, the reader should check if the researcher has performed a rigorous estimation of the population standard deviation, either empirically or theoretically. The reader can also research herself if the population standard deviation is reasonable, possibly estimate it herself.

2.2.4. *Changing a Two-tailed Test into a One-tailed Test*

So far we examined one-tailed tests, in which we have one direction of interest, therefore testing if the result is either lower or higher

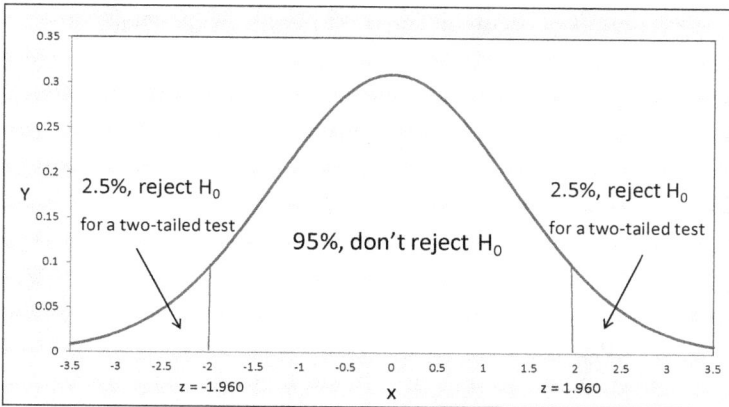

Figure 2.19: Normal distribution with a 5% significance level for a two-tailed test, and the resulting critical z-scores of $z = -1.960$ and $z = 1.960$.

than a benchmark. So in a one-tailed test the alternative hypothesis typically has a bigger or smaller sign. We discussed the one-tailed test examples of a hedge fund manager having a significantly higher return than the market in Example 2.1, and if Jeff is significantly taller than the US male average in Example 2.2.

However, our interest may be bi-directional. We could test if a medication leads to a higher life expectancy for some patients *and* a lower life expectancy for others. Or we can ask if people after marrying are happier or more frustrated. It's probably both ☺. So in a two-tailed test, the alternative hypothesis typically has an $\neq$ sign.

Why is this important? Well, in a two-tailed test we have to look into both tails. By convention, for a 5% significance level, each tail now represents 2.5% of the distribution. Therefore the critical z-values are lower for the left tail and higher for the right tail compared to a one-tailed test. Indeed they are -1.960 for the left tail and 1.960 for the right tail. In Excel, we derive this as $=$ normsinv(0.025) $= -1.960$ and $=$ normsinv(0.975) $= 1.960$, see Appendix A.4 for details. A spreadsheet that derives z-scores from a standard normal distribution can be found at www.dersoft.com/normsinv.xlsx. Figure 2.19 shows the critical regions of a two-tailed test.

As we can see from Figure 2.19, the total rejection percentage is still 5%, however, now split up into two 2.5% tails.

So how can we cheat with this? Well, it is easier to achieve statistical significance with a one-tailed test since the area of significance is concentrated in one tail! In other words, the critical z-scores of a one-tailed test are lower, $1.645 < 1.960$ in the right tail, and the same for the left tail, just in absolute terms $|-1.645| < |-1.960|$. Therefore the critical z-scores in a one-tailed test are easier to beat, meaning statistical significance is easier to achieve.

So a cheater can change an original bi-directional research hypothesis into a one-directional hypothesis. For our Example 2.1 of the hedge fund manager Mr. Wonderful, the original alternative hypothesis may have been if he overperforms *or underperforms* the market. Mr. Wonderful's z-score was 1.649. As we showed, this is statistically significant for a one-tailed test, since $1.649 > 1.645$ (the latter being the z-score to beat in a one-tailed test). However, in a two-tailed test Mr. Wonderful's performance is not statistically significant, since his z-score is smaller than the critical two-tailed z-score, i.e., $1.649 < 1.960$. The same logic applies to our Example 2.2 where Jeff is statistically taller than the US male in a one-tailed test, but not a two-tailed test.

Somewhat more serious can be changing a bi-directional test into a one-directional test in medicine. If we want to know if a medication has positive effects on some patients as well as negative effects, a bi-directional test is sensible. Changing this bi-directional test into a one-directional one, may inflate statistical significance with respect to efficacy, and additionally ignore negative side effects!

How to catch a cheater who changes the original bi-directional hypothesis into a one-directional one? Well, the reader should observe if the researcher has analyzed and justified his choice of the one-tailed test and then conclude if the choice is reasonable. The reader can also do her own analysis and use common sense to derive if a one-tailed test is sensible. Generally, be aware of one-tailed tests! Statistical significance is often easily achieved, particularly in combination with the degree of freedom to estimate the population mean μ and the population standard deviation σ, as discussed in Sections 2.2.2 and 2.2.3.

**Summary and Conclusion: How to Catch Inference
Cheaters — Piece of Cake?**

Statistical Inference tries to determine if an association is occurring
by change or if it is significant. For example, we want to know if a
drug has a significant impact on an illness, or if piano players have
more luck with girls. From my personal experience, the latter is not
the case, but I am only a sample of one ☺.

Statistical significance is typically determined with the z-score.
The equation is

$$z = \frac{\bar{x} - \mu}{\sigma/\sqrt{n}}, \tag{2.16}$$

where $\bar{x}$ is the sample mean, μ is the (typically unknown) population
mean, σ is the (typically unknown) population standard deviation,
and n is the sample size.

The z-score concept has weaknesses, which can be exploited by
a cheater in numerous ways:

(1) A cheater can increase the sample size n, for example increase the
number of people in a study on the efficacy of a medication, or
increase the number of days in a study of a mutual fund manager
claiming to significantly outperform the market. Increasing the
sample size n, increases the z-score as we can see (or at least I,
sorry for the arrogance ☺) from Equation (2.16)). So the bigger
the sample size n, the more likely that the z-score is statistically
significant.

(2) A cheater can also manipulate the typically unknown population
mean μ. As we can see from Equation (2.16), the lower the
estimate μ, the higher the z-score and the more likely that the
z-score is statistically significant.

(3) To increase significance, it is easiest to manipulate the typically
unknown population standard deviation σ, also called "nuisance
parameter", since it often is, due to lack of data, difficult to
estimate. From Equation (2.16), we can observe that the lower
the estimation for σ, the higher the z-score, i.e., the higher the
likelihood to achieve a statistically significant outcome.

A further way to achieve statistical significance is to change an originally bi-directional hypothesis into a single-directional hypothesis. For example, in a bi-directional study, we are interested if a drug has a positive impact on an illness *and* a negative one due to the side effects. Or we want to know if a tax reduction leads to more government revenue due to more economic activity as Ronald Reagan in the 1980s, Kansas governor Sam Brownback in 2012 and other politicians have falsely claimed. At the same time we want to know if tax cuts simply reduce government revenue, which has been the case in the past.

A bi-directional hypothesis has two areas of significance, one in the extreme left and one in the extreme right tail (see Figure 2.19). Changing the bi-directional hypothesis into a single-directional one, reduces the area of significance into *one* tail (see Figures 2.17 and 2.18). Therefore, the critical z-score is lower (for the left tail lower in absolute terms) and statistical significance is easier to achieve.

How to Catch Inference Cheaters? Not Totally Easy

With respect to the sample size n, it has to be "reasonable". A too small sample size means little statistical rigor, a too large sample size inflates inference results. The reader may check if the researcher has justified his choice of sample size, do her own reasonability study on the sample size, and use common sense.

With respect to estimating the population mean μ, often some data is available to verify the researcher's estimation (for example the mean height of US citizens is available). The researcher's estimation of the population standard deviation σ is more difficult to assess. So again, the reader may check if the researcher has justified his estimation, either theoretically or empirically or do her own analysis.

Changing a bi-directional hypothesis into a one-directional one increases the probability of significance since the area of significance is in a single tail and therefore larger. The problem is that the hypothesis may not be sensible anymore since an important aspect of the analysis may be lost. For example, only the positive effects of a drug are studied, but not the negative side effects. The reader should use common sense to validate if a one-directional study is sensible.

Questions and Problems

The answers are available to instructors, please email gunter@ dersoft.com.

1. What are the two critical parameters of the Pearson correlation model? What is their relationship?
2. What does the correlation coefficient ρ tell us?
3. What does the regression coefficient β tell us?
4. Explain the nonlinearity drawback of the Pearson correlation model! What is the remedy?
5. The linearity limitation of the Pearson model is addressed with nonlinear fitting. What is Underfitting?
6. What is Overfitting?
7. If earlier data is considered as relevant as more recent data, what polynomial is considered a good fit?
8. If later data is considered as relevant as more recent data, what polynomial is considered a good fit?
9. What is spurious correlation, also called correlation $\neq$ causation?
10. Give two examples of spurious correlation.
11. When two variables, for example swimming pool accidents and heart attacks, are indirectly correlated because they are correlated to a third variable, for example a heat wave, is this nonsense correlation?
12. What do we mean by Kitchen Sink Regression?
13. What is a good remedy for the Kitchen Sink Regression problem?
14. What is multicollinearity? Why is it a problem? What is the remedy?
15. What is heteroskedasticity? Why is it a problem? What is the remedy?
16. What does Inferential Statistics try to accomplish?
17. The z-score equation is $z = \frac{\bar{x} - \mu}{\sigma/\sqrt{n}}$. Explain the variables.
18. How can a statistical cheater cheat with the sample size n? What is the remedy?

19. How can a statistical cheater cheat with the population mean μ? What is the remedy?
20. Why is population standard deviation σ called a "nuisance parameter"? How can a statistical cheater cheat with it? What is the remedy?
21. Why is statistical significance easier achieved with a single-tailed test?
22. What is the problem when changing a two-tailed test hypothesis into a single-tail hypothesis?

Appendix A.1: Nasty Math Calculations for the Standard Deviation, Covariance, and the Correlation Coefficient

Ok, for all those who want to know the tedious math to calculate regression parameters, here it is:

From the data in Table A.1, we derive the standard deviation of the returns of asset X as

$$\sigma_X = \sqrt{\frac{1}{n-1} \sum_{t=1}^{n} (X_t - \overline{X})^2}$$

$$= \sqrt{\frac{1}{4} \sum_{t=1}^{n} \begin{array}{l} (0.4055 - 0.2059)^2 + (-0.1823 - 0.2059)^2 \\ +(0.1823 - 0.2059)^2 + (0.0645 - 0.2059)^2 \\ +(0.5596 - 0.2059)^2 \end{array}}$$

$$= 0.2899.$$

Table A.1: Prices of assets X and Y and their returns.

Year	Asset X	Asset Y	Returns x (%)	Returns y (%)
2015	$100	$200		
2016	$150	$270	40.55	30.01
2017	$125	$460	−18.23	53.28
2018	$150	$410	18.23	−11.51
2019	$160	$480	6.45	15.76
2020	$280	$380	55.96	−23.36
		Average =	**20.59**	**12.84**

The same exercise for the returns of asset Y, results in

$$
\sigma_Y = \sqrt{\frac{1}{n-1} \sum_{t=1}^{n} (Y_t - \overline{Y})^2}
$$

$$
= \sqrt{\frac{1}{4} \sum_{t=1}^{n} \begin{array}{l} (0.3001 - 0.1284)^2 + (0.5328 - 0.1284)^2 \\ +(-0.1151 - 0.1284)^2 + (0.1576 - 0.1284)^2 \\ +(-0.2336 - 0.1284)^2 \end{array}}
$$

$$
= 0.3099.
$$

Following Equation (2.3), the covariance of X and Y is

$$
\text{COV}_{XY} = \frac{1}{n-1} \sum_{t=1}^{n} (x_t - \overline{x})(y_t - \overline{y})
$$

$$
= \frac{1}{4}[(0.4055 - 0.2059)x(0.3001 - 0.1284)
$$

$$
+ (-0.1823 - 0.2059)x(0.5328 - 0.1284)
$$

$$
+ (0.1823 - 0.2059)x(-0.1151 - 0.1284)
$$

$$
+ (0.0645 - 0.2059)x(0.1576 - 0.1284)
$$

$$
+ (0.5596 - 0.2059)x(-0.2336 - 0.1284)] = -0.0623.
$$

See www.dersoft.com/Pearsonmodel.xlsx for the calculations in Excel.

Appendix A.2: On the Slope of the Regression function

In this appendix, we will show why the slopes of reciprocal regression functions are not reciprocal. Important stuff! Let's revisit Figure 2.14.

OK, the slope of the regression function in Figure 2.14 is $\hat{\beta}_1$. It shows how much Y changes if X changes by an infinitesimally small amount. We write this $\hat{\beta}_1 = \frac{dY}{dX}$. For practical purposes, we can approximate the change in X, dX, by 1. So if $\hat{\beta}_1$ would be 0.5,

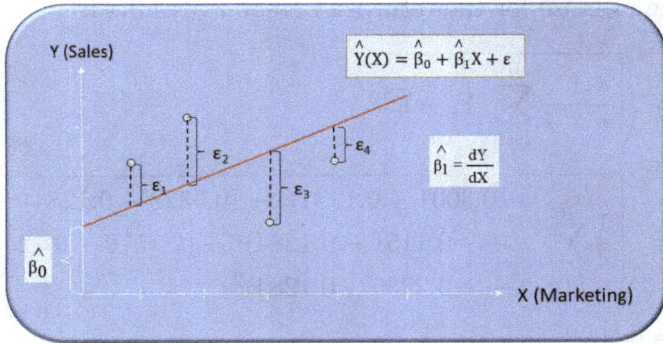

Figure 2.14: Four given data points and the resulting regression function

$$\hat{Y}(X) = \hat{\beta}_0 + \hat{\beta}_1 X + \varepsilon. \tag{A.1}$$

this means that for every \$1 more spent on Marketing, on average, Sales increases by 50 cents.

Now assume we run the reciprocal regression

$$\hat{X}(Y) = \hat{\alpha}_0 + \hat{\alpha}_1 Y. \tag{A.2}$$

So we would analyze if a change in Sales Y has an impact on Marketing X. Makes some sense: If Sales increase, more money may be allocated to Marketing. Now you would assume that the slopes of the regression Functions (A.1) and (A.2) are reciprocal, i.e., $\hat{\beta}_1 = \frac{1}{\hat{\alpha}_1}$. However, this is typically not the case. True is that in most cases we have

$$\hat{\beta}_1 \neq \frac{1}{\hat{\alpha}_1}. \tag{A.3}$$

Why is this so? Easy: In the regression analysis $\hat{Y}(X) = \hat{\beta}_0 + \hat{\beta}_1 X + \varepsilon$, we minimize the *vertical* errors, i.e., the vertical differences between the data points and the regression function as seen in Figure A.1. In the regression analysis $\hat{X}(Y) = \hat{\alpha}_0 + \hat{\alpha}_1 Y + \varepsilon$, we minimize the *horizontal* errors, i.e., the horizontal differences between the data points and the regression function. So the two analyses minimize different distances, which makes the slopes of the regression functions non-reciprocal. For a numerical example see, www.dersoft.com/nonsensecorrelation.xlsx, cells M21 vs S22, and cells M22 vs S21.

When do we actually have $\hat{\beta}_1 = \frac{1}{\hat{\alpha}_1}$? It is when the correlation coefficient $\rho = 1$ and the resulting regression function has a slope of 1.

Appendix A.3: Derivation of SST = SSE + SSR

A key objective in regression analysis is explaining the value of a variable or its change, the latter we call variation (not to be confused with variance, the square of standard deviation, see Equation (2.2)). For example, we want to explain life expectancy Y with healthy lifestyle X. However, only part of life expectancy can be explained with healthy lifestyle since there are other factors such as genetics or access to health care which influence life expectancy.

So only part of Y can typically be explained with one regressor X. We quantify this explained part by SSE, the explained sum of squares. We use squares to get rid of nasty negative numbers. And part of Y cannot be explained with X, which we quantify with SSR, residual sum of squares. And we can sum up the two, to get the total variation, which we call SST. So we have

$$\text{SST}_{\text{otal}} = \text{SSE}_{\text{xplained}} + \text{SSR}_{\text{esidual,}} \qquad (\text{A.4})$$

in more detail, $\quad \sum (y_i - \overline{y})^2 = \sum (\hat{y}_i - \overline{y})^2 + \sum (y_i - \hat{y}_i)^2, \quad (\text{A.5})$

where SS is the sum of squares, SST is the total sum of squares, SSE is the explained sum of squares, SSR is the residual sum of squares, y_i are the y-axis values of the given data points, $\overline{y}$ is the average of the y_i, and $\hat{y}_i$ are the y-axis values on the regression function.

A word of caution: On the web the reader will sometimes find SSE and SSR defined oppositely as here, so SSE would be the residual sum of squares (our SSR), and SSR would be the explained sum of the squares (our SSE). Shame on you statisticians for using different notations! ⊗ I am a statistician but I don't include myself in this mess...

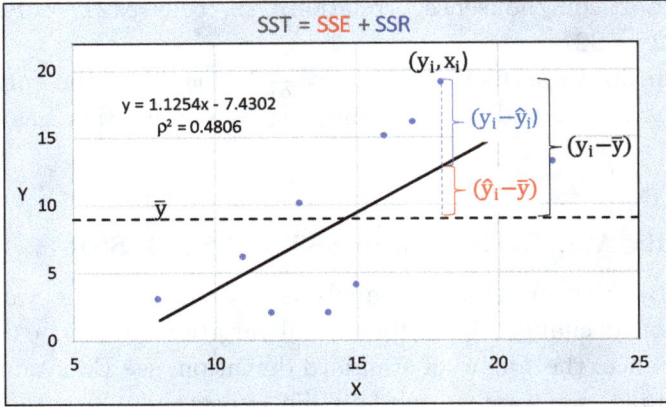

Figure A.1: Total variation of Y, $(y_i - \overline{y}) =$ Explained variation of Y, $(\hat{y}_i - \overline{y}) +$ unexplained variation of Y, $(y_i - \hat{y}_i)$ for a certain data point (y_i, x_i).

A proof of Equation (A.5) is at www.dersoft.com/sstproof.pdf, but it is probably easier to understand the concept of Equations (A.4) and (A.5) by just looking at Figure A.1.

Figure A.1 can be found at www.dersoft.com/sst.xlsx. OK, the basic work is done. Now let's show that increasing the number of regressors X typically increases the correlation coefficient ρ or the coefficient of determination ρ^2.

ρ^2 is defined as the fraction of the sample variation of Y that is explained by X. So we can write

$$\rho^2 = \frac{\text{SSE}}{\text{SST}} = \frac{1 - \text{SSR}}{\text{SST}}. \tag{A.6}$$

Expanding Equation (2.9), we write the multivariate regression function as

$$\hat{Y}(X) = \hat{\beta}_0 + \hat{\beta}_1 X_1 + \hat{\beta}_2 X_2 + \cdots + \hat{\beta}_n X_n + \varepsilon. \tag{2.9a}$$

Now with every new regressor X that is added to the regression model, a new term to Equation (2.9a) (graphically a new axis) is added, which adds explanatory power to the model, i.e., increases SSE, unless the correlation coefficient ρ and the regression coefficient $\hat{\beta}_1$ of the new regressor are both zero. However, when a new regressor X is added, the total sum of squares SST remains unchanged, since

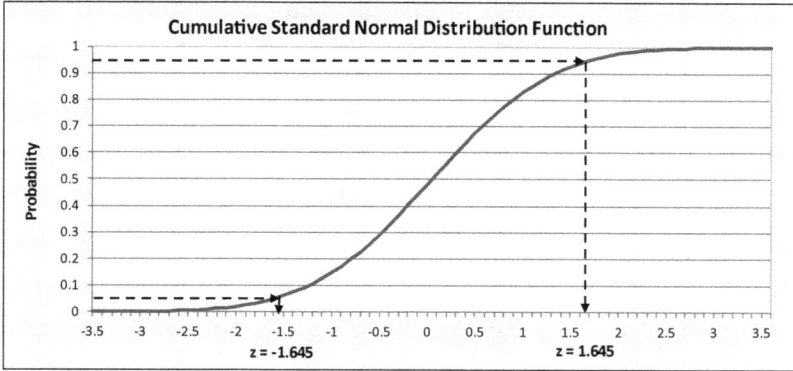

Figure A.2: Derivation of the z-score for a 5% significance level.

SST is only impacted by y-axis coordinates and not by x-axis values. Since SSE increases and SST stays constant, from Equation (A.6), ρ^2 increases, q.e.d.

Appendix A.4: Derivation of the z-Score

In math, we typically have an independent variable X, which is mapped with a function f to a dependent variable Y, so we write $Y = f(X)$. For example, we have $Y = X^2$. To derive the z-score we use the opposite procedure: Y is given and then mapped with the cumulative normal distribution function to X. That's why deriving the z-score is also called "inversion".

A picture is worth 1,000 words (and an equation is worth 1,000 pictures, however, not everyone agrees with that ☺) so let's explain the derivation of the z-score graphically.

From Figure A.2, we can observe that the z-score is derived in two steps:

(1) A significance level is determined, which is a probability, displayed on the Y-axis. In Figure A.2, the significance level is 5%.
(2) For the left tail, we use the probability of 0.05 and move horizontally to the right until a point on the cumulative distribution function is reached. We then move down to find the x-axis value, which is the z-score of $z = -1.645$. In other words,

the probability of 0.05 is mapped via the cumulative normal distribution to a z-score value.

To find the z-score for the right tail, we apply the same procedure. We now use a probability of 0.95, move horizontally to the right to find a point on the cumulative distribution, then move downwards to find the x-axis value, which is the z-score of $z = 1.645$. So the probability of 0.95 is again mapped via the cumulative normal distribution to a z-score value.

In this way, we can naturally find the z-score for any significance level.

References

Meissner, G., *Correlation Risk Modeling and Management*, 2nd edition, RISKBOOKS, 2019.

Vasicek, O., *Probability of Loss on a Loan Portfolio*. KMV Working Paper, 1987 (Results published in *RISK* magazine with the title "Loan Portfolio Value", December 2002).

Chapter 3

"Creative Reporting", i.e., Distorting Outputs

"Figures will not lie, but liars will figure"

(Carrol D. Wright)

In this chapter, we will show how to mislead a reader about the statistical results. There are two main ways to achieve this honorable objective:

(1) Creating a Numerical Bias, which means reporting numbers which are incomplete, flawed, or incorrect.
(2) Creating a Visual Bias, so distorting output graphs via image distortion, truncating axes, different axis scales, omitting data, or displaying a 3D problem misleadingly as 2D.

3.1. Numerical Bias

There are different ways to mislead a reader about numerical results. One is using the property that returns, i.e., percentage changes are not additive. This can generate profits, which do not exist! Let's show this.

3.1.1. *Profits from Nothing: Adding Percentage Changes*

In finance, growth rates are expressed as percentage changes, which we calculate with equation

$$\frac{S_t - S_{t-1}}{S_{t-1}}, \tag{3.1}$$

where S_t and S_{t-1} are the prices of an asset at time t and $t-1$, respectively.

For example, if $S_{t-1} = 100$ and $S_t = 110$, the percentage change is $\frac{110-100}{100} = 0.1 = 10\%$.

The problem is that these percentage changes are not additive in time! Let's show this with an example:

Example 3.1: *Let's assume we have stock price moving as displayed in Figure 3.1.*

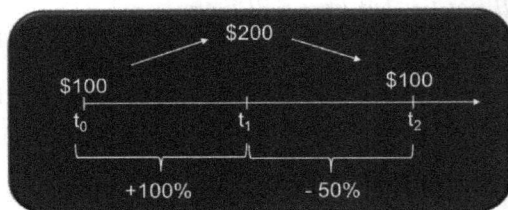

Figure 3.1: Stock price movement and its percentage changes.

In Figure 3.1, we have a stock price in t_0 of $100. The stock moves to $200 in t_1 and back to $100 in t_2. From Equation (3.1), the percentage change from t_0 to t_1 is ($200 − $100)/$100 = 100\%. The percentage change from t_1 to t_2 is ($100 − $200)/$200 = −50\%. Adding the percentage changes, from t_0 to t_2 we derive +100% − 50% = +50%, although the stock has not increased from t_0 to t_2! What on earth is going on? Well, the way we defined the percentage changes in Equation (3.1), the percentage changes are not additive. In other words, a cheating mutual fund can generate profits, which do not exist!

What is the solution to this problem? Logarithmic changes:

We often approximate percentage changes with the help of the natural logarithm ln:

$$\frac{S_t - S_{t-1}}{S_{t-1}} \approx \ln\left(\frac{S_t}{S_{t-1}}\right). \tag{3.2}$$

This is a good approximation for small differences between S_t and S_{t-1}. For example, if $S_{t-1} = 100$ and $S_t = 101$, from Equation (3.2) we have

$$\frac{101 - 100}{100} \approx \ln\left(\frac{101}{100}\right)$$

or $\quad 1\% \approx 0.9950\%.$

$\ln(S_t/S_{t-1})$ is called a *log-return*. The advantage of using log-returns is that they can be added over time. Let's show this with our Example 3.1:

The log-return from t_0 to t_1 is $\ln(200/100) = 69.31\%$. The log-return from t_1 to t_2 is $\ln(100/200) = -69.31\%$. Adding these log-returns in time, we get the correct return of the stock price from t_0 to t_2 of $69.31\% - 69.31\% = 0\%$.

What is the drawback of using logarithmic changes? Well, the logarithmic approximation of percentage changes of Equation (3.2) becomes imprecise for big changes. We just saw that in our example: A stock price increase from \$100 to \$200, so an increase of 100\%, leads to a logarithmic increase of 69.31\%. A mutual fund telling the investor that a stock has increased by 69.13\% when it has increased by 100\% is naturally quite confusing ☹.

In conclusion, the reader should be aware of a cheater who adds percentage changes. This can lead to incorrect, inflated results! A remedy is using logarithmic changes. However, the drawback is that logarithmic changes become imprecise for single periods if the change of the asset price is big. The best solution is to take the starting value and the ending value and calculate the percentage change (with Equation (3.1)). A spreadsheet showing percentage changes and logarithmic changes and their additions in time, is at www.dersoft.com/additivity.xlsx.

The non-additivity of percentage changes is one way to cheat or mislead on stock or portfolio performance. How to ethically measure and present portfolio performance has become a big issue in finance. There are the Global Investment Performance Standards (GIPS). Investment firms often try to generate confidence in their company by advertising that they are "GIPS compliant". And if that's not enough, you can get a certification on the topic, the Certificate in Investment Performance Measurement (CIPM), for more see https://www.investopedia.com/terms/c/certificate-invest ment-performance-measurement-cipm.asp.

3.1.2. *Reporting Bias in Academia*

In this section, we will explain how statistical results can be misreported. We will look at two types: Reporting bias in academia, and reporting bias in practice. Generally, we can define

> *Reporting Bias is selective revealing or suppression of statistical results*

Let's start with reporting biases in academia.

PhD students are in the rat race. They are struggling to get a professor position at a university. How to win rate race? Publications! The quantity of publications is important, but more important is the quality of the journal where the paper is published. How to get into a good journal? Well, besides the quality of the paper, a statistically significant result is highly beneficial. So if a researcher finds that a drug can cure cancer, or that CO_2 emissions lead to global warming, or a neural network can statistically significantly forecast stock prices, these are promising results. In fact, statistically significant results are published about 3 times more often than non-significant results. This is also called the *publication bias* or *file drawer bias*, since many papers with a non-significant result end up in a file drawer.

The fact that it is promising to find statistically significant results, inspires some researchers to manipulate their study, so that the study generates statistically significant outcomes. This can be done in three ways:

(1) Manipulate the input data, as we explained in Chapter 1.

Table 3.1: R&D and revenue (in million).

R&D	Revenue
$5.60	$10.00
$5.50	$100.00
$5.00	$100.00
$4.50	$70.00
$4.50	$50.00
$6.00	$90.00
$5.50	$30.00
$5.00	$35.00
$6.00	$60.00

(2) Manipulate the statistical calculations as we explained in Chapter 2.

(3) Manipulate the outputs, which we will now discuss.

Selective Significance Reporting: Only Reporting Statistically Significant Parameter Values

Example 3.2: *PhD student Disearnest is looking for a job in R&D (Research and Development). He wants to show that R&D has a positive impact on the Revenue of companies. He collects the above data in Table 3.1:*

Graphically, Table 3.1 is displayed in Figure 3.2.

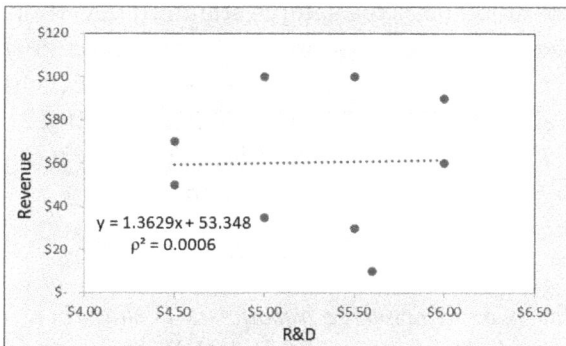

Figure 3.2: Graphical display of R&D and revenue of Table 3.1.

From Figure 3.2, we already assume that there is little correlation between R&D and Revenue: The data is distributed in quite a circular way.

Disearnest calculates the correlation coefficient ρ between R&D and Revenue. It is 0.0241. Disappointing! The coefficient of determination ρ^2 is naturally even more disappointing. It is 0.0006 as seen on Figure 3.2 (the reader can also find these values at www.dersoft.com/ t-test.xlsx, sheet "Rho vs Beta", cells c15 and c16). As we discussed in Appendix A.3 of Chapter 2, ρ^2 can be interpreted as the fraction of the sample variation of Y that is explained by X. So only 0.06% of the Revenue is explained by R&D!

We can also run a t-test to find that R&D is not a significant factor in explaining Revenue. A t-test is very similar to the z-test which we explained in Section 2.2 of Chapter 2 with Equation (2.16). The equation for the t-test is

$$t = \frac{\overline{x} - \mu}{s/\sqrt{n}}, \tag{2.16a}$$

where $\overline{x}, \mu$, and n are defined as in the z-test: $\overline{x}$ is the sample mean, μ is the (typically unknown) population mean, n is the sample size, however, the (typically unknown) population standard deviation σ is replaced with the sample standard deviation s.

We use the t-test instead of the z-test if the sample size is small, typically for $n < 30$ and if the sample standard deviation s is known, which it typically is. If the t-score is higher than a critical positive value or lower than a critical negative value, we conclude that the test result is statistically significant (see Figure 2.19).

For our Example 3.2, our null-hypothesis is that there is no correlation between R&D and Revenue, so we have

$$H_0 : \rho = 0.$$

Our research or alternative hypothesis is that there is a positive or negative correlation between R&D and Revenue. So we conduct a two-tailed test:

$$H_1 : \rho \neq 0.$$

As derived above, the sample correlation coefficient $\rho = 0.0241$. In Equation (2.16a), we set μ to zero, since we do not want to know the difference between $\bar{x}$ and the population correlation coefficient μ, but the difference between $\bar{x}$ and zero. For a 95% confidence level, assuming that the sample standard deviation of ρ is 10%, and we have $n = 9$ data points, we get from Equation (2.16a):

$$t = \frac{\rho - 0}{s/\sqrt{n}} = \frac{0.0241 - 0}{0.1/\sqrt{9}} = 0.7230.$$

The critical 95% t-value for a two-tailed test can be found in statistical tables, which you can use if you are born before 1980 ☺, sorry for the age discrimination. But is easier to use software. In Excel, we use $= T.INV.2T(1\text{-}0.95, 9\text{-}1\text{-}1) = 2.3646$ (see cell S5 in file www.dersoft.com/t-test.xlsx). In Python, we can use

```
INPUT
import scipy.stats
sig = 0.05 #significance level
df = 7 #degrees of freedom since n-k-1 = 9-1-1 = 7
print(scipy.stats.t.ppf(1-sig/2,df))

OUTPUT
2.364622510102993
```

The user can run this code in the cloud at https://replit.com/@Gun terMeissner/t-test#main.py.

Since $0.7230 < 2.3646$, we accept the null-hypothesis that there is no statistically significant impact of R&D on Revenue (which we had already assumed by looking at Figure 3.2).

Frustrated by the statistically insignificant correlation coefficient ρ, Disearnest calculates the regression coefficient β. From Equation (2.5), it is

$$\beta = \frac{COV_{XY}}{\sigma_X^2} = \frac{0.4444}{0.3261} = 1.3629.$$

The reader can find this value at www.dersoft.com/t-test.xlsx, sheet "t-test", cell B26. This means that for every increase in R&D of $1, Revenue increases on average by $1.36. That's more like it! Earnest calculates if β is statistically significant: Following Equation (2.16a), the t-test is now

$$t = \frac{\beta - \beta_P}{s/\sqrt{n}}, \tag{2.16b}$$

where β is the slope of the sample regression function, β_P is the slope of population regression function, s is the standard deviation of β, and n is the sample size.

The null-hypothesis is that the slope of Revenue with respect to R&D is zero:

$$H_0 : \beta = 0.$$

Our research or alternative hypothesis is that the slope is unequal zero. So we conduct a two-tailed test:

$$H_1 : \beta \neq 0.$$

The sample regression coefficient β (derived above) is $\beta = 1.3629$. We set β_P to zero, since we do not want to know the difference between β and the population regression coefficient β_P, but the difference between β and zero. For a 95% confidence level, assuming that the standard deviation of β is 60%, we get from Equation (2.16b):

$$t = \frac{\beta - \beta_P}{s/\sqrt{n}} = \frac{1.3629 - 0}{0.6/\sqrt{9}} = 6.8145$$

The reader can find this value at www.dersoft.com/t-test.xlsx, sheet "t-test", cell G5.

The critical 95% t-value for a two-tailed test (as derived above) is 2.3646. Since $6.8145 > 2.3646$, the regression coefficient β of Revenue with respect to R&D is statistically significant! (see also Figure 2.18). Disearnest shows his statistically significant regression coefficient β to his company of choice and does not report the statistically insignificant correlation coefficient ρ. Will he get the job? No. The *company looks at his data, calculates the statistically insignificant correlation coefficient ρ, unmasks his cheating and tells him to go home.*

In conclusion, in statistical tests it can happen that some outcome coefficients are statistically significant, and others are not. A cheater can do "selective significance reporting", i.e., only report significant coefficients. Is this actually statistical cheating? Well yes. Critical information is deliberately hidden, which constitutes cheating!

In our Example 3.2, why actually is the correlation coefficient ρ statistically insignificant, and the regression coefficient β significant? Easy. We discussed the ρ, β relationship in Equation (2.6)

$$\beta = \rho \frac{\sigma_Y}{\sigma_X}. \tag{2.6}$$

From Table 3.1, the standard deviation of Y (revenue), σ_Y is 32.2533. However, the standard deviation of X (R&D), σ_X is only 0.5711 (the reader can find these values at www.dersoft.com/t-test.x lsx, sheet "Rho vs Beta", cells C20 and C21). So from Equation (2.6), with $\rho = 0.0241$ (derived above), we have

$$\beta = \rho \frac{\sigma_Y}{\sigma_X} = 0.0241 x \frac{32.2533}{0.5711} = 1.3629.$$

Vice versa, if we have data for X and Y values with the standard deviation of X much larger than the standard deviation of Y, we can derive the opposite result, i.e., that ρ is statistically significant and β is not. I am sure the reader can't wait to do this exercise with problems 6 and 7 at the end of this chapter ☺.

One more word about Figure 3.2. The slope of the regression function is 1.3629. So the regression function grows with an angle of $\tan^{-1}(1.3629)$.[1] We can use a calculator or Excel to find =Degrees(atan(1.3629)) = 53.73°. In Python, we can use

---

[1]Tan stands for the tangent function. To find the angle x of the slope 1.3629, we use $\tan(x) = 1.3629$. To solve for x, we multiply both side with the tangent inverse $\tan^{-1}$. Using $\tan^{-1}(\tan(x)) = x$, we get $x = \tan^{-1}(1.3629)$. This can be calculated in Excel with = Degrees(atan(1.3629))= 53.73°. ("atan" is the syntax for $\tan^{-1}$ in Excel).

```
INPUT
import math
print(math.degrees(math.atan(1.3629)))
OUTPUT
53.73140258873423
```

(code can be run at https://replit.com/@GunterMeissner/Tangent-inverse#main.py).

My point is that the slope of the regression function in Figure 3.2 looks close to zero, not 1.3629! the regression function increases with an angle of 53.73°, higher than 45°, which does not seem to be the case in Figure 3.2! What happened? Well, we have different units on the Y-axis and X-axis: The units are larger on the X-axis than on the Y-axis. Therefore the regression function is visually compressed. We will discuss graphs with misleading units in detail in Section 3.2.2.

> *"Mutual Funds invest your money until its gone"*
>
> (Woody Allen)

3.1.3. *Reporting Bias in Practice*

In this section, we will discuss reporting biases, mainly using examples in economics and finance. There are several types of reporting biases. Let's start with two examples where output data is suppressed.

Reporting Bias by Arbitrary Data Suppression

Let's first look at some basic terms. A Mutual Fund is a company that pools capital from investors and invests it for them. How have mutual funds performed in the past? Lousy ☺. About 90% of mutual funds underperform the market! This is due to two main reasons: (1) Financial markets have become quite "efficient", meaning there are few overvalued or undervalued stocks; and (2) The mutual fund managers need to be paid, so there are fees.

A Fund of Funds is a mutual fund who invests in other mutual funds. This is basically a good idea, since the degree of diversification is high. Let's explain the reporting bias with a fund of funds.

Table 3.2: Performance of
10 mutual funds in 2021
ranked by performance

Fund	Performance
1	27%
2	20%
3	12%
4	8%
5	5%
6	4%
7	3%
8	1%
9	−5%
10	−13%

Example 3.3: *The fund of funds Malevolent has invested in 10 funds, which have performed in 2021 as displayed in Table 3.2.*

The average performance of all 10 funds is

$$(27\% + 20\% + 12\% + 8\% + 5\% + 4\% + 3\% + 1\% - 5\%$$
$$- 13\%)/10 = 6.20\%$$

The fund of funds is quite disappointed with the performance. It decides to eliminate funds 9 and 10 from its portfolio. This means that the performance is

$$(27\% + 20\% + 12\% + 8\% + 5\% + 4\% + 3\% + 1\%)/8 = 10.00\%.$$

Going forward, the Malevolent fund will only invest in funds 1 to 8. So it reports a performance of 10.00% for 2021. Is this statistical cheating? Of course. The historical performance in 2021 is 6.20%, not 10.00%. It is a classical case of reporting bias!

How high is the reporting bias? Really easy. We can use equation

$$\text{Reporting bias} = P_s - P_a, \tag{3.3}$$

where P_s is the performance of the selected (arbitrarily reduced) fund, and P_a is the performance of all funds in the portfolio. So in our example, the Reporting bias = 10.00% − 6.20% = 3.80%.

How to catch a reporting bias by data suppression? Well, mutual fund positions and their change are typically publicly available, so we can check if data is eliminated. We can also look at audit reports, which try to verify the validity of the reporting. This has gone wrong in the past though, as the famous Enron scandal in 2001 with $71 billion in hidden losses, the WorldCom scandal in 2002 with $11 billion inflated assets, and the Bernie Madoff Ponzi scheme with about $40 billion in non-reported losses show. We discuss the Madoff Ponzi scheme in more detail below.

Survivorship Bias

Survivorship bias is another form of a reporting bias where data is suppressed. However, whereas in the previous section the data suppression was arbitrary, in a survivorship bias data is suppressed since companies have defaulted or merged. Before we look at an example, let's first explain what a hedge fund is. Well, just like a mutual fund, a hedge fund pools capital from investors and invests it for them. However, hedge funds invest in riskier assets (for example junk bonds), use modern technology like high frequency trading and algorithmic trading, and use derivatives such as futures, swaps, and options. Therefore, hedge funds are riskier and sometimes blow up. Let's look at a survivorship bias example.

Example 3.4: *Let's assume a hedge fund index is comprised of 100 equally weighted hedge funds, with an average return in year 2021 of $+8\%$. Three of the hedge funds, which had a return of -30%, -50% and -70%, blew up, and were eliminated from the index. If the index is reported without the three defaulted hedge funds, what is the return R, i.e., the survivorship bias? We dig deep into our 6th grade algebra and write*

$$97R + (3 \times -50\%) = 100 \times 8\%.$$

Solving for R gives us $R = 9.79\%$. Therefore the survivorship bias is

$$9.79\% - 8\% = 1.79\%.$$

Is reporting the 9.79% return cheating? Yes. The 2021 return was 8%, not 9.79%, so the return is falsely overstated due to the survivorship bias!

How to catch survivorship cheaters? Quite easy: Data of hedge funds who defaulted and their performance is typically available. So we can calculate, just as we did in Example 3.4, which hedge fund defaulted and what impact the default has on the index.

Backfill Bias

In the previous two sections, biased reporting was achieved by data suppression. Let's now discuss the opposite: Biased reporting by data addition. This is called backfill biases, inclusion bias or instant history bias. Let me explain the backfill bias with another example.

Example 3.5: *The hedge fund "Greed is good" (a phrase from the famous movie Wall Street) had 8 positions in its portfolio, which have performed in 2021 as displayed in Table 3.3.*

A position may simply be a stock or a bond, or an option, future, or swap. The average performance of the eight positions is

$$(17\% - 10.2\% + 12\% - 3\% - 35\% + 22\% + 8.9\%$$
$$+ 4.3\%)/8 = 2.00\%.$$

Unhappy with the mediocre performance of 2.00%, the hedge adds two positions to its portfolio. Position 9, let's say a stock, has

Table 3.3: Return of positions of Hedge Fund "Greed is Good".

Position	Performance
1	17.0%
2	−10.2%
3	12.0%
4	−3.0%
5	−35.0%
6	22.0%
7	8.9%
8	4.3%

Table 3.4: New portfolio of
Hedge Fund "Greed is Good"
with added positions 9 and 10.

Position	Performance
1	17.0%
2	−10.2%
3	12.0%
4	−3.0%
5	−35.0%
6	22.0%
7	8.9%
8	4.3%
9	16%
10	12%

performed +16%, and position 10, let's assume a junk bond, has
performed +12%. So the portfolio now is as shown as Table 3.4.

The hedge fund will have these 10 positions starting in 2022.
Including the 10 positions, the average performance of the 10 positions in 2021 is

$$(17\% - 10.2\% + 12\% - 3\% - 35\% + 22\% + 8.9\% + 4.3\% + 16\%$$
$$+ 12\%)/10 = 4.40\%.$$

However, the positions 9 and 10 were not part of the portfolio in
2021! They are "filled back" into the portfolio for the year 2021 and
the hedge fund reports a return of 4.40% for 2021. It is cheating? Of
course. What is the backfill bias? Very easy. We use equation

$$Backfill\ bias = P_b - P_o, \tag{3.4}$$

where P_b is the performance of the portfolio with the backfilled
positions, and P_o is the performance of the original (true) portfolio.
So for our example we have

$$Backfill\ bias = 4.40\% - 2.00\% = 2.40\%.$$

How to catch a backfill cheater? Well, we can try to get data
from the hedge fund and see when and which positions were added

to the portfolio. This is typically not easy since most hedge funds are quite secretive and not obligated to display their positions and performance.

"*Die Gedanken sind frei*" ≡ "*There is freedom in thought*"

Ponzi Scheme

Naturally, there are many other ways of "Creating Reporting". An infamous one is the Ponzi scheme, perfected by the legendary Bernie Madoff. He managed to overstate earnings of about $40 billion over 15 years. How did he do it? Simple: He paid out investors by contributions of new investors. This works as long as the capital from new investors is higher than the payout to leaving investors. To fuel the contributions of new investors, naturally, Bernie Madoff reported consistently miraculous earnings with little risk (i.e. low volatility) over many years. What caused the blow-up of the Madoff Ponzi Scheme? It was the 2008 global financial crisis: Many investors tried to withdraw their investments, which could not be met by new investors.

In 2009, Madoff pleaded guilty to 11 felonies, including securities fraud, wire fraud, mail fraud, money laundering, making false statements, and perjury. Madoff claimed he was solely responsible for the scheme. However, hiding losses of $40 billion over 15 years can't be done by a single individual. Several of his employees in accounting and employees of the auditing firm Friehling & Horowitz CPAs received long prison sentences. While pleading guilty, Madoff did not plea bargain with authorities, possibly to avoid exposing family members involved in the scheme. He was sentenced to 150 years in prison and died of natural causes in 2021.

How to catch a Ponzi–Schemer? It's not so easy. Madoff was able to fool auditors for 15 years! Red flags, which investors should watch out for, are as follows:

(1) Consistently high returns, even in a financial crisis. Madoff reported positive returns even in the global financial crisis in 2008.

(2) Complex strategies which may hide losses.

(3) Investments that are not registered with the SEC and can therefore be hard to audit.

(4) Investments from feeder funds (small funds that give their portfolios to bigger funds to achieve economies of scale) which add auditing complexity.

(5) Difficulty to get payouts.

Here are two more simple ways of creative reporting.

Time Frame Reporting Bias

One simple way of misreporting is stating a misleading reporting time frame. Let's say a fund exists for 15 years. It had bad performances in years 3, 4, and 5. The fund now simply reports the good performance of the last 10 years.

Fund Reporting Bias

Mutual funds and hedge funds typically run several individual funds, for example an international fund, a high-risk fund with start-up companies, or just a bond fund. Naturally some of the funds perform better than others. A mutual or hedge fund can be creative in just reporting the ones that have performed well. Is this cheating? Well yes. Critical information is deliberately suppressed, which constitutes cheating.

3.2. Visual Bias

After having discussed numerical reporting biases, let's now look at ways to mislead a reader with distorted images and graphs. Let's start with a simple image manipulation.

3.2.1. *Image Distortion*

Example 3.6: *The anti-environmental think tank "Planet last" disputes deforestation. In fact, they found certain regions in which*

Figure 3.3: Image increase.

Figure 3.4: The image increase of Figure 3.3 is 9-fold.

forestation from 2000 to 2020 has increased 3-fold. Planet last displays this graphically in Figure 3.4.

What is wrong with Figure 3.4? Well, the image increase is not 3-fold, but 9-fold, since not only the height, but also the width is amplified by a factor 3.

So in conclusion, never trust anti-environmentalists...

Ok, let's get a bit more serious and look at sophisticated graph distortions. We can define:

> *A distorted graph is an intentional or unintentional misrepresentation of data in a graph which may lead to wrong conclusions*

An easy way to achieve a distorted graph is manipulating the axes. Let's look at some examples.

3.2.2. *Truncating the Y-Axis*

Example 3.7: *The start-up Apocalypse and Sons is not doing well. The annual revenue from 2016 to 2020 is displayed in Table 3.5, column 2.*

Table 3.5: Revenue of Apocalypse and Sons, % changes, and GDI.

Year	Revenue	%change of revenue, displayed in figure 3.5(a)	%change from distorted figure 3.5(b)	Graph distorsion index GDI
2016	100,000			
2017	101,000	1.00%	50.00%	50.00
2018	102,000	0.99%	33.33%	33.67
2019	103,000	0.98%	25.00%	25.50
2020	104,000	0.97%	20.00%	20.60

The revenue of column 2 with a zero origin of the Y-axis is graphically displayed in Figure 3.5(a).

The management of Apocalypse is unhappy with Figure 3.5(a) since the increase in revenue can hardly be noticed. So it creates Figure 3.5(b) with a 98,000 origin of the Y-axis. The data in Figures 3.5(a) and 3.5(b) are identical, just the origin of the Y-axis is different!

What is the correct display, Figure 3.5(a) or 3.5(b)? We can argue that a zero origin displayed in Figure 3.5(a) is a reasonable reference point, which displays the small increases in revenue of 1% and less each year, see Table 3.5, column 3, in a sensible way.

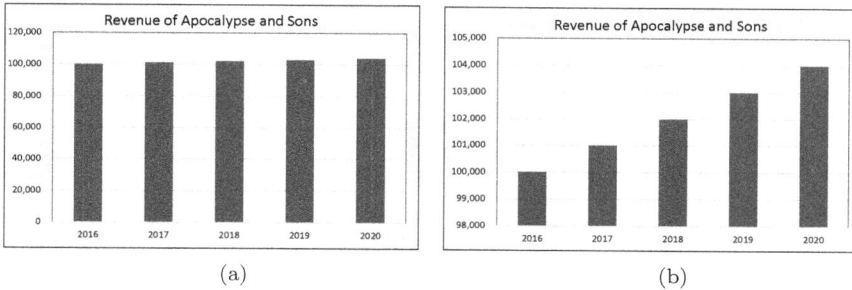

(a)

(b)

Figure 3.5: (a) Revenue of Table 3.5 with a zero origin of the Y-axis. (b) Revenue of Table 3.5 with a 98,000 origin of the Y-axis.

Figure 3.5(b) is visually misleading. Visually the revenue increase from 2016 to 2017 is 50%, from 2017 to 2018 33%, from 2018 to 2019 25%, and from 2019 to 2020 20%, see column 4 in Table 3.5. A spreadsheet with these data can be found at www.dersoft.com/truncatedaxis.xlsx.

How much is Figure 3.5(b) misleading? We can use the simple equation

$$GDI = a/b, \qquad (3.5)$$

Where GDI is the Graph Distortion Index, a is the %change in distorted graph and b is the %change in data.

For our example, GDI is displayed in the last column of Table 3.5. We see that Figure 3.5(b) is extremely misleading: For 2017, a 1% increase in revenue is displayed as a 50% increase in Figure 3.5(b)! So the GDI for 2017 is 50 or 5,000%! A GDI of ±5% is often considered a benchmark for graphical distortion. Why actually is Figure 3.5(b) extremely distorted? The truncation of the Y-axis is very high (98,000) in comparison to the increases in the data, which is small (see column 2 in Table 3.5).

So is truncating the Y-axis graphical cheating? If it results in a significant distortion, which is typically assumed if the GDI is ±5%, then yes. So, the reader should be aware of truncation cheaters, and not be misled by truncated Y-axes!

3.2.3. *Different Axes Units*

In Figure 3.5, the X and Y axis variables were measured on a different scale. The X-axis displayed time, the Y-axis US dollar. If the X-axis and the Y-axis are measured on the same scale, for example both are measured in Celsius, meters, or US dollar, it is sensible to have the same units on the X and Y-axis. If not, this can be misleading as we already saw with Figure 3.2: Since the X-axis units were much larger than the Y-axis units, the regression function was compressed and visually displayed with a lower slope than the actual slope. Let's discuss this in more depths with an example.

Example 3.8: *The marketing team of Sus LLC (Sus is Zoomer abbreviation for suspicious. I have two kids and therefore speak Zoomer ☺) was quite successful in recent years. Table 3.6 shows the association between marketing and sales.*

Running a regression for the data in Table 3.6, we realize that for every dollar spent on Marketing, Sales have increased on average by a dollar, as displayed in Figure 3.6.

Figure 3.6 displays the data in a sensible way. The units on the X-axis are the same as the units on the Y-axis. The slope of the regression function is 1.00, see the equation on Figure 3.6, or the reader can look at www.dersoft.com/differentaxisunits.xlsx, cell D20.

Table 3.6: Marketing and sales of Sus LLC, in $US

	Marketing	Sales
	X	Y
2010	4,520	4,300
2011	4,649	5,600
2012	6,500	7,625
2013	5,184	4,500
2014	5,208	6,635
2015	5,757	6,000
2016	5,500	6,500
2017	5,896	6,800
2018	6,381	6,500
2019	6,933	7,000
2020	7,362	7,728

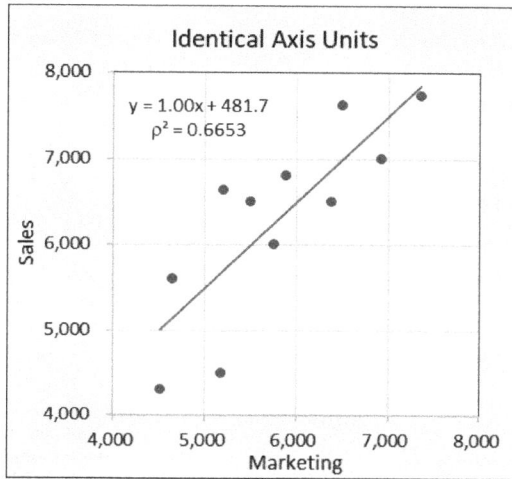

Figure 3.6: Marketing and sales association of Sus LLC for data in Table 3.6. Units on the Y-axis and X-axis are identical.

As we can visually observe, the slope of 1.00 (a 45° angle) is graphically correctly displayed in Figure 3.6. Therefore, there is no distortion of the graph.

Despite the decent relationship between marketing and sales, the Sus marketing team wants to impress the board, and derives Figures 3.7(a) and 3.7(b).

In Figures 3.7(a) and 3.7(b), we can see that the regression function is displayed with a slope of 2, although the data in Table 3.6 results in a slope of 1 (see www.dersoft.com/differentaxisunits.xlsx, cell D20). Is this cheating? Why yes! Is it a deliberate visual misrepresentation of the data! Shame on you Sus marketing team!

Luckily Jim from IT, a passionate marketing foe, has access to the data in Table 3.6. He creates Figure 3.8 to discredit the marketing team:

From Figure 3.6, we observe that the slope of the regression function is now 0.5, despite the data in Table 3.6, which results in a slope of 1. So Jim is also a statistical cheater!

In conclusion, if the X-axis and Y-axis are measured on the same scale, (for example both in inches, Celsius or dollars), it is sensible to display the X-axis and the Y-axis in the same units. Otherwise the graph can be visually misleading and can lead to false conclusions!

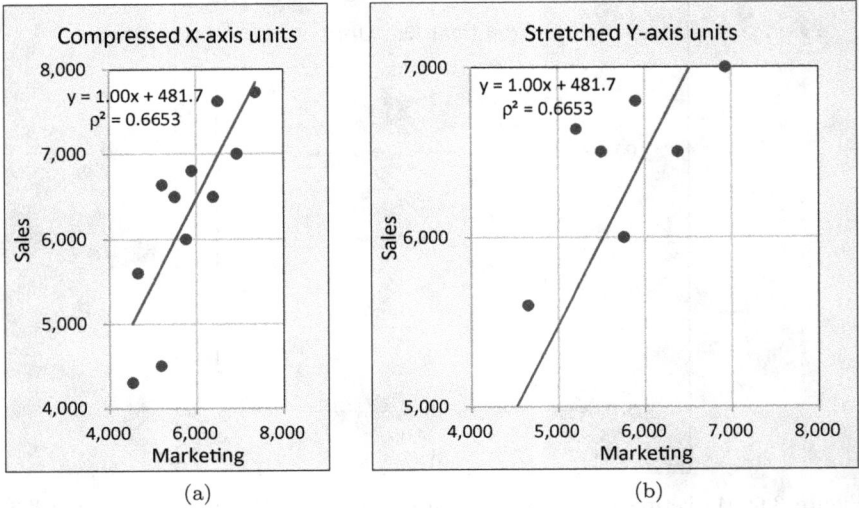

Figure 3.7: (a) Data of Table 3.6. X-axis units are compressed. (b) Data of Table 3.6. Y-axis units are stretched.

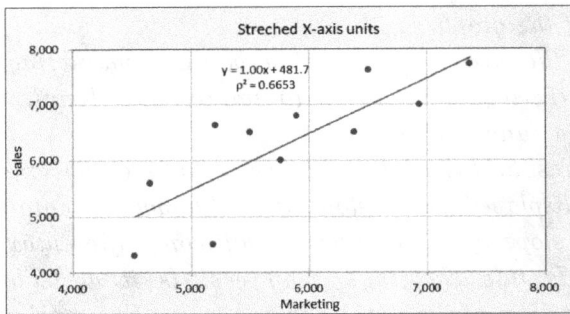

Figure 3.8: Data from Table 3.6. X-axis units are stretched.

3.2.4. *Logarithmic Display*

In science, we quite often display the Y-axis on a logarithmic scale. Let's have a look if this statistically legitimate.

Figure 3.9 shows the Dow Jones Industrial Average from 1920 to 2021.

From Figure 3.9, we observe that the Dow Jones has been quite volatile, however in the long run has increased exponentially:

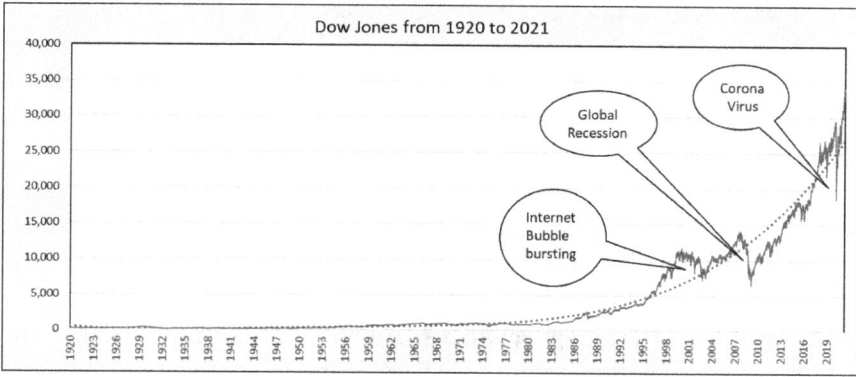

Figure 3.9: The Dow Jones Industrial Average on a standard (non-logarithmic) scale. The polynomial (dotted line) is of order 4.

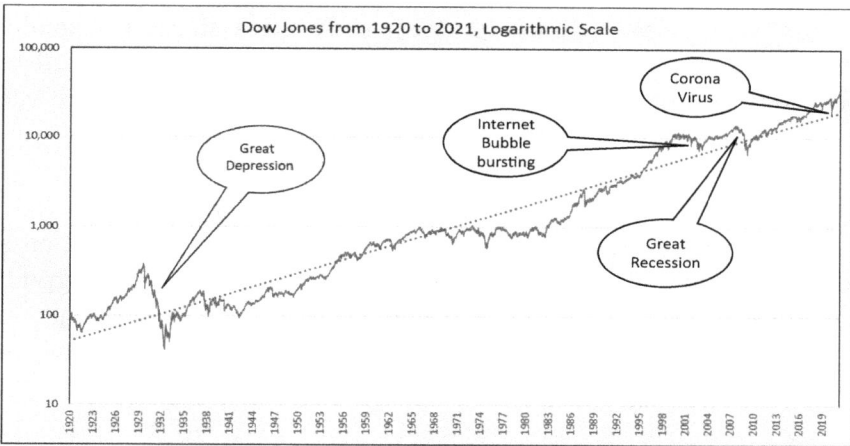

Figure 3.10: Dow Jones Industrial Average from 1920 to 2021. Y-axis displayed on a logarithmic scale. The polynomial fit is of order 1, so it is linear.

The exponential polynomial (dashed line) fits the data well. So stockholders should invest in the long run and not complain (which I do not). From Figure 3.9, we can also observe that the difference in low numerical data from 1920 to about 1980 can hardly be seen. That's where the logarithmic display of Figure 3.10 comes in.

What are the take-aways from Figure 3.10? Well, we see that low numerical data can now be clearly viewed. We can see the decline in the Dow Jones during the Great Depression from 1929 to 1932. The drawback however is that differences in the high numerical values are more difficult to see. We can hardly observe the Internet Bubble bursting in 2001, the Great Recession 2008 and Corona Virus crash in 2020.

In conclusion, a logarithmic scale shows the percentage changes of the data. This can be very informative: In the Great Depression, the Dow Jones declined by 90%!, during the Internet Bubble bursting 20.4%, the Great Recession 54.1% and the Corona Virus crash 37%. These critical percentage changes are visible in Figure 3.10. The attentive reader will remember that we discussed percentage changes, which are used in the volatility calculation, see Appendix A.3 of Chapter 1, and also in Section 3.1.1.

Some technical stuff: The logarithmic display of the Y-axis in Figure 3.10 is to the base 10. We can see this since the equally distanced units are in multiples of 10. So the distances from 10 to 100, 100 to 1,000, 1,000 to 10,000 and 10,000 to 100,000 are the same.

So, is displaying the Y-axis logarithmically statistical cheating? No. However, a researcher should clearly state that the y-axis (and sometimes also the X-axis, called a log-log graph) is displayed on a logarithmic scale. Otherwise, the conclusion drawn from the graph can be misleading. For example, the increase of the Dow Jones in Figure 3.10 appears to be linear, however it is exponential as seen in Figure 3.9.

3.2.5. *Omitting Data*

In Section 1.1.5, we discussed the volatility concept. It measures how much a variable like a stock price fluctuates. For the math see Appendix A.3 of Chapter 1.

In finance, volatility is a measure of risk. The idea is: the more a stock price fluctuates, the riskier it is, i.e., the more it can decline. Does that make sense? Basically yes. However, if a stock price

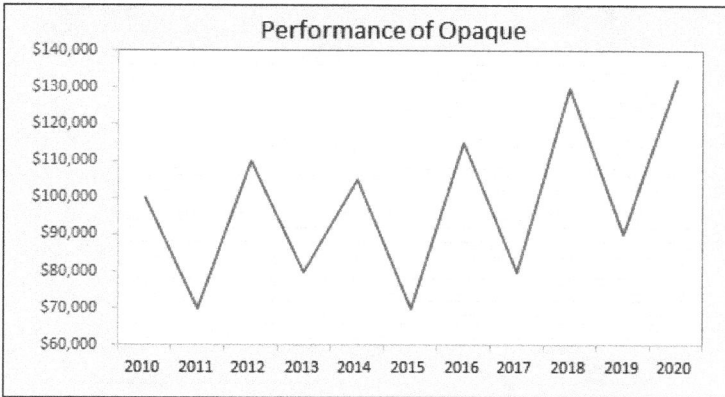

Figure 3.11: Annual performance of Opaque from 2010 to 2020.

fluctuates a lot, this means it can also increase a lot! So savvy investors like me ☺, like high volatility stocks!

However, most investors don't like volatility. Volatility can be hidden in a graph if certain data points are omitted. Let's look at an example:

Example 3.9: *In the last 10 years, the hedge fund "Opaque" had some good years and bad years as displayed in Figure 3.11.*

Opaque wants to convince its investors that the performance fluctuation is low. Therefore, they omit the performance of uneven years between 2010 to 2020, so for the years 2011, 2013, 2015, 2017, and 2019. This results in Figure 3.12.

The objective is achieved, the performance in Figure 3.12 is less volatile than the performance displayed in Figure 3.11. In fact, the volatility of the performance in Figure 3.11 is 41.62% and the performance volatility in Figure 3.12 is 6.96%, which the reader can verify at www.dersoft.com/dataomission.xlsx, cells C18 and C28.

Is data omission in figures statistical cheating? Sure. Critical information is deliberately omitted, so the graphical output is deliberately misleading!

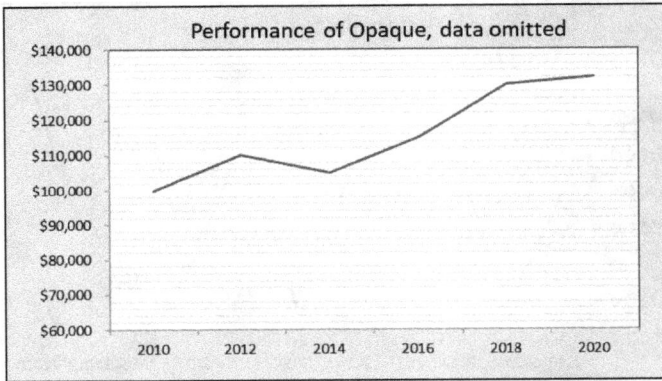

Figure 3.12: Bi-annual performance of Opaque from 2010 and 2020.

3.2.6. *2D Versus 3D Data Representation*

Ok, I created the last example to prevent math geeks from complaining that this book is trivial!

Some concepts cannot be meaningfully displayed in a 2D graph. However, a 3D graph may be sufficient. The Gumbel copula[2] is a good example. It is displayed in 2D in Figure 3.13.

It seems that in Figure 3.13 most data points lie close to the origin (the area 0,0). However, if we display the Gumbel copula in 3D, we see a different result.

From Figure 3.14, we observe that most data points lie in the upper right corner, the area close to (1,1). Figure 3.13 cannot display this property since it only has 2 dimensions, the width x and the height y, but it lacks the depths dimension c_{GU} (which stands for Gumbel copula). Simply put, in Figure 3.13 the data points close to

---

[2]The Copula function was transferred from statistics to finance in 2000 by David Li. It was enthusiastically embraced and presumably able to correlate the 125 assets in a Collateralized Debt Obligation (CDO). The copula function became infamous in the 2008 global financial crisis and was wrongfully blamed for it. For details see Meissner (2019), *Correlation Risk Modeling and Management*, second edition, Chapter 6.

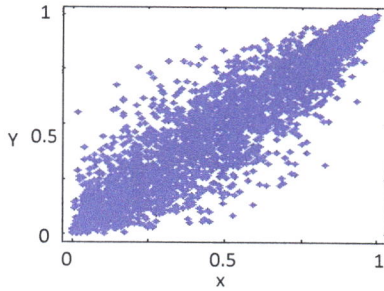

Figure 3.13: Gumbel Copula in 2D.

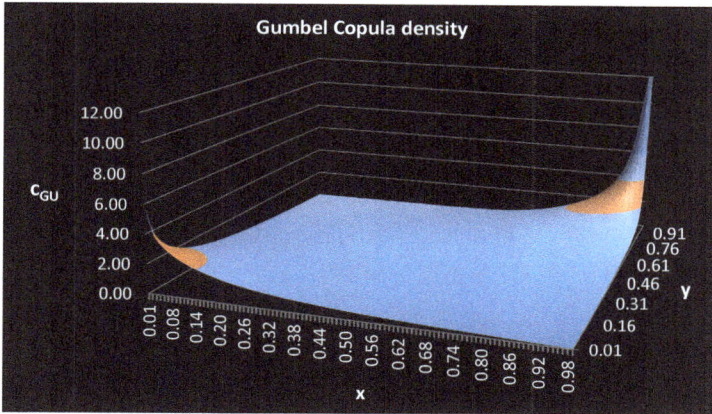

Figure 3.14: Gumbel Copula in 3D.

(1,1) "lie on top of each other", which cannot be observed in the 2D display.

In conclusion, some concepts are too complex to be displayed in a 2D graph. A 3D graph may reveal the critical properties. And naturally some concepts are too complex to be displayed in a graph at all. This is typically the case when there are more than 3 variables involved. For example, our multiple regression analysis of Equation (2.9),

$$\hat{Y}(X) = \hat{\beta}_0 + \hat{\beta}_1 X_1 + \hat{\beta}_2 X_2 + \hat{\beta}_3 X_3 + \varepsilon, \qquad (2.9)$$

has 3 regressors X and one dependent variable $\hat{Y}(X)$ and cannot be graphically displayed. So we have to understand it without the help of graphs. But as already stated in this book: An equation is worth 1,000 pictures, so we are all good ☺. The math and a model of the Gumbel copula (and Gaussian copula) can be found at www.dersoft.com/Gumbelcopula.xlsx.

Summary: How to Identify Creative Reporters — No Problem

In this chapter, we discussed how cheaters can manipulate the statistical output. This can be achieved by

(1) manipulating the output numbers; and
(2) manipulating the output graphs.

(1) With respect to manipulating the numbers, one blunt way is to add percentage changes. This can lead to profits which do not exist! So if a mutual fund really adds percentage changes, send them Chapter 3!

In academia, another way of reporting manipulation is selective reporting, so just reporting statistical outputs which are favorable. A researcher may only report output parameters which prove that a hypothesis is statistically significant. This typically helps to get a paper published. So the reader, stats savvy after reading this book, should watch for missing statistical outputs.

In financial practice, a cheater may also perform selective reporting. For example, a mutual fund only reports the returns of favorable years, or the returns of certain profitable funds.

In addition, hedge fund reporting is often enhanced by the "survivorship bias": Only those hedge funds who did not default report, inflating the hedge fund performance. Popular is also the "backfill bias": New positions or funds are included into the report. Their excellent performance is then "filled back" into the report, although the positions or funds were not part of the portfolio. Mutual fund performance is audited, publicly available and can be checked. Hedge funds however are less strictly regulated and quite secretive, so their data is more difficult to verify.

Another way of misreporting results is the famous Ponzi scheme, perfected for 15 years by Bernie Madoff. He simply paid out leaving investors with capital from new investors. This works as long as the incoming capital from the new investors is higher than the outflow of leaving investors. Madoff got caught during the global financial crisis 2008, when many investors requested payouts, which could not be met by new capital. And we gratefully acknowledge: In the end, cheaters get caught! For five warning signs of a possible Ponzi scheme, see Section 3.1.2.

(2) Manipulating output graphs is another way of "creative reporting". Many manipulations of graphs are possible. Truncating the y-axis can make differences, for example in revenue growth, seem large when they are not. If the units of the x-axis and y-axis are identical, for example, they are both in meters, kilogram, or Celsius (yes, the European scales are way more practical ☺), it is sensible to have the same axis units. If not, a relationship between X and Y can be distorted. For example, a higher increase in Y (for example of profits) can be misleadingly displayed, or a lower increase (for example CO_2 emissions) can be displayed.

In economics and finance, we often rescale the exponential growth of a variable, for example GDP or stock prices, on a logarithmic scale. This creates a less steep, often linear function. Is this statistical cheating? No. It actually reveals valuable information, since the graph now displays percentage changes. However, the logarithmic scaling should be pointed out by the researcher, so no false conclusions are drawn.

Simply omitting data in the graphical display is another way to mislead the audience. It can lead to higher return averages and less volatility. Since volatility is often interpreted as risk, risk-adjusted performance measures can be inflated.

Some concepts can be too complex to be displayed meaningfully in a 2D graph. However, a 3D graph may be able to reveal critical information of the concept.

Questions and Problems

The answers are available to instructors, please email gunter@derso ft.com

1. Create a numerical example which shows that percentage changes are not additive in time.
2. What is the functional relationship between the correlation coefficient ρ and the regression coefficient β?
3. What condition has to be fulfilled for ρ to be statistically significant but not β?
4. What condition has to be fulfilled for β to be statistically significant but not ρ?
5. If the regression coefficient $\beta = 0.1$, the standard deviation of x, $\sigma_X = 20$ and the standard deviation of Y, $\sigma_Y = 0.25$, what is the correlation coefficient ρ? (Use Equation (2.6)).
6. Let's assume $\rho = 0.8$, the sample size $n = 9$, we have only one regressor k, and the standard deviation of ρ is 50%. Is ρ statistically different from 0 on a 95% confidence level?
7. Let's assume $\beta = 0.1$, the sample size $n = 9$, we have only one regressor k, and the standard deviation of β is 50%. Is β statistically different from 0 on a 95% confidence level?
8. Give an example of "arbitrary data suppression" in reporting.
9. What is the survivorship bias and what is the consequence?
10. What is a Ponzi scheme? What is the condition for a Ponzi scheme to work?
11. What is the backfill bias and what is the consequence?
12. Give an example of an image distortion.
13. Give an example of truncating the y-axis. What can be the result?
14. Why is it not a good idea to display different axes units, if the variables are measured on the same scale?

15. What does the logarithmic display accomplish?
16. Is the logarithmic display of data statistical cheating?
17. Give an example how omitting data can alter a graphical result.
18. 2D graphs may not be able to meaningfully display a problem. Why not?